INTELLECTUAL PROPERTY
AND GENETICALLY MODIFIED ORGANISMS

Charles Lawson – for CP and Victoria

Berris Charnley – for Marija, 'Death doesn't Bargain.'

Intellectual Property and Genetically Modified Organisms

A Convergence in Laws

Edited by

CHARLES LAWSON
Griffith Law School, Griffith University, Australia

BERRIS CHARNLEY
St Anne's College, University of Oxford, United Kingdom

Routledge
Taylor & Francis Group

LONDON AND NEW YORK

First published 2015 by Ashgate Publishing

Published 2016 by Routledge
2 Park Square, Milton Park, Abingdon, Oxon OX14 4RN
711 Third Avenue, New York, NY 10017, USA

First issued in paperback 2017

Routledge is an imprint of the Taylor & Francis Group, an informa business

British Library Cataloguing in Publication Data
A catalogue record for this book is available from the British Library

The Library of Congress has cataloged the printed edition as follows:
Intellectual property and genetically modified organisms : a convergence in laws / by Charles Lawson and Berris Charnley.
    p. cm.
  Includes bibliographical references and index.
  ISBN 978-1-4724-4345-8 (hardback)
  1. Genetically modified foods--Law and legislation. 2. Intellectual
property. 3. Plants, Cultivated--Patents. I. Lawson, Charles, editor. II. Charnley, Berris, editor.
  K3927.I58 2015
  346.04'84--dc23

                                                                2014031727

ISBN 13: 978-1-138-08852-8 (pbk)
ISBN 13: 978-1-4724-4345-8 (hbk)

Contents

List of Tables

List of Cases

Notes on Contributors

Berris Charnley is a Postdoctoral Research Assistant at St. Anne's College, University of Oxford. In 2008 he helped establish the IPBio Network (ipbio.org), an international group of scholars with a shared research interest in the history of IP in the biosciences and its relevance to current policy. Berris' research encompasses the history and philosophy of science (and in particular biology, plant breeding, genetics and genomics); the history and theory of intellectual property; and the history of food science.

Stephen Hubicki is a PhD Candidate with the Department of History and Philosophy at the University of New South Wales, and an Adjunct Research Fellow at the Australian Centre for Intellectual Property in Agriculture, Griffith Law School, at Griffith University. Stephen's current research examines the history of patenting medical discoveries in Britain, and the treatment of intellectual property in wartime.

Fran Humphries is a PhD Candidate with the Australian Centre for Intellectual Property in Agriculture, Griffith Law School at Griffith University. Fran's research focuses on aquaculture and its exploitation with a particular emphasis on intellectual property.

Charles Lawson is a Professor at Griffith University and a member of the Australian Centre for Intellectual Property in Agriculture, Griffith Law School at Griffith University. Charles' research interests span intellectual property, competition, trade, access and benefit-sharing, and public administration.

Karinne Ludlow is a Senior Lecturer at Monash University. Karinne recently co-authored a report for the Australian Government on future-proofing regulations for new science and technology – *Regulatory Models for Future Developments in Enabling Technologies*. Karinne researches interdisciplinary problems of science and technology and the law, with particular emphasis on biotechnology and nanotechnology.

Dianne Nicol is a Professor at the University of Tasmania. Dianne's research tracks the regulation of biotechnology and human genetics and the commercialisation of genetic knowledge and patenting genetic inventions. She is currently the lead investigator on two Australian Research Council funded projects, one on the role of material transfer agreements in exchanges of biological materials, the other on the role of law in the era of personalised medicine.

Matthew Rimmer is an Australian Research Council Future Fellow, an Associate Professor at the ANU College of Law and an Associate Director of the Australian Centre for Intellectual Property in Agriculture at the Australian National University. Matthew's research focuses on intellectual property and climate change and his broader interests include copyright law and information technology; patent law and biotechnology; access to medicines; clean technologies and traditional knowledge.

Jay Sanderson is a Lecturer at Griffith University and a member of the Australian Centre for Intellectual Property in Agriculture, Griffith Law School at Griffith University. Jay researches intellectual property with an emphasis on the relations between plants, science, politics, people and practice.

Kieran Tranter is a Senior Lecturer at Griffith University and a member of the Socio-Legal Research Centre, Griffith Law School at Griffith University. Kieran researches and writes on the jurisprudential and cultural dimensions of law and technology.

Acknowledgements

This collection was instigated by our desire to address the question of what has happened to intellectual property in the context of genetically modified organisms (GMOs), and what this relationship can tell us about intellectual property at large. While GMOs were apparently peripheral to our current research interests a superficial analysis showed that actually, intellectual property and GMOs have become enmeshed in very interesting ways. Stepping back from the subject revealed that analysis of relations between intellectual property and GMOs provides interesting insights into several perennial intellectual property debates as well as some of the most recent developments. Discussions among our colleagues encouraged us to try to bring some of these ideas together. In bringing this all together we acknowledge the generous financial support of the Australian Centre for Intellectual Property in Agriculture, the Australian Research Council's *Discovery Project* funding scheme (DP120101434), and the Griffith Social and Behavioural Research College. We also acknowledge the discussion, insight, support and assistance of our colleagues, especially Brad Sherman, Carol Ballard and Julia Barker, that has made this book possible.

List of Abbreviations

AHTEG	Ad Hoc Technical Expert Group
AIPPI	Association Internationale pour la Protection de la Propriété Intellectuelle
AMP SCOTUS	*AMP v. Myriad Genetics*
AOC	French Appellation of Control
ASTA	American Seed Traders Association
BCH	Biocultural heritage
BCHI	BCH Innovation
BIO	Biotechnology Industry Organization
BRCA genes	Two human genes which had been shown to be linked with hereditary forms of breast cancer
BSA	Business Software Association
CAT	Chloramphenicol acetyltransferase
CBD	Convention on Biological Diversity
cDNA	complementary DNA
CFR	Code of Federal Regulations
CFSAN	Center for Food Safety and Applied Nutrition
CJEU	Court of Justice of the European Union
CPB	Cartagena Protocol on Biosafety
CSIRO	Commonwealth Scientific and Industrial Research Organisation
CVM	Centre for Veterinary Medicine
DNA	Deoxyribonucleic acid
EPA	Environmental Protection Agency
EPC	European Patent Convention
EPSPS	5-enolpyruvylshikimate-3-phosphate synthase
ESC	Economic and Social Committee
EU	European Union
FDA	Food and Drug Administration
FD&C Act	Federal Food, Drug and Cosmetic Act 1938
gDNA	genomic DNA
GE	Genetically engineered
GI	Geographic indications
GM	Genetically modified
GMOs	Genetically modified organisms

Guidance 187	Guidance for Industry 187: Regulation of Genetically Engineered Animals Containing Heritable DNA Constructs
IEF	Isoelectric Focusing
IGC	WIPO's Intergovernmental Committee on Intellectual Property and Genetic Resources, Traditional Knowledge and Folklore
IIED	International Institute for Environment and Development
INPI	Instituto Nacional de la Propiedad Industrial Argentina
Mayo SCOTUS	*Mayo Collaborative Services v. Prometheus Laboratories, Inc.*
NIAB	National Institute of Agricultural Botany
op-AFT	Ocean Pout Antifreeze protein
PCPs	polychlorinated biphenyls
PCR	Polymerase chain reaction
rDNA	recombinant deoxyribonucleic acid
RFE	FDA's Regulatory Fish Encyclopaedia
RR	Roundup Ready
SCOTUS	Supreme Court of the United States
SDNY	Southern District of New York
SIFOR	smallholder innovation for resilience project
SPRC	Seafood Products Research Center
TK	Traditional knowledge
TPP	Trans-Pacific Partnership Agreement
TTIP	Transatlantic Trade and Investment Partnership
TAFTA	Transatlantic Free Trade Agreement
TRIPS	Agreement on Trade-Related Aspects of Intellectual Property Rights
UPOV Convention	International Union for the Protection of New Varieties of Plants
UN	United Nations
US	United States
USC	United States Code
USDA	US Department of Agriculture
USPTO	United States Patent and Trademark Office
VCM	Center for Veterinary Medicine
VMAC	Veterinary Medicine Advisory Committee
WG	Working Group
WIPO	World Intellectual Property Organisation
WTO	World Trade Organisation

Chapter 1

Intellectual Property
and Genetically Modified Organisms

Berris Charnley and Charles Lawson

The first modern genetically modified organisms (GMOs) appeared on the planet in the early 1980s. Public debates about these astonishing new objects were fierce. They centred mainly on the politics of regulating risk. In the years since those first debates agricultural (green) and medical (red) GMOs have entered the marketplace in several countries. Inevitably, many areas of law have played a role in this process as GMOs are controlled to varying extents, and in several different ways, from their inception in the laboratory all the way through to their use by consumers. This entanglement of GMOs and law has evolved into a rich area of research. While the politics of regulating the risks of GMOs remain an active and incomplete project, it is the other areas of law in this entanglement that have captured our attention. Of particular interest is the convergence of GMOs with the broadly conceived area of intellectual property.

This volume sits at the intersection of two developing research fields. On the one hand the extensive literature on GMOs is increasingly coming to consider issues beyond risk. The prospective tone of early classics on regulation, such as Michael Ruse and David Castle's 2002 *Genetically Modified Foods: Debating Biotechnology,*[1] has been replaced with a more formalised tone, embodied in works such as Luc Bodiguel and Michael Cardwell's 2010 *The Regulation of Genetically Modified Organisms: Comparative Approaches,*[2] or Victor Tutelyan's, *Genetically Modified Food Sources: Safety Assessment and Control.*[3] Authors interested in GMO have increasingly begun to publish on a broader range of issues, for example, Gary Marchant, Guy Cardineau and Thomas Redick in their 2010 edited volume, *Thwarting Consumer Choice: The Case against Mandatory Labelling for Genetically Modified Foods*[4] move from the traditional debate of in-field

---

1   Ruse, M. and Castle, D. 2002. *Genetically Modified Foods: Debating Biotechnology.* Amherst, NY; Prometheus Books.

2   Bodiguel, L. and Cardwell, M. 2010. *The Regulation of Genetically Modified Organisms: Comparative Approaches.* Oxford, UK: Oxford University Press.

3   Tutelyan, V. 2013. *Genetically Modified Food Sources: Safety Assessment and Control.* London, UK; Academic Press.

4   Marchant, G., Cardineau, G. and Redick, T. (eds) 2010. *Thwarting Consumer Choice: The Case against Mandatory Labelling for Genetically Modified Foods.* Washington, DC; Government Institutes.

regulation to a consideration of in-store regulation. As scholarship on GMOs has changed focus in the last two decades, so has scholarly analysis of intellectual property. Once considered a strange and peripheral area of law, recent volumes, such as Mario Biagioli, Peter Jaszi, and Martha Woodmansee's 2011 *Making and Unmaking Intellectual Property: Creative Production in Legal and Cultural Perspective*[5] and Alain Pottage and Brad Sherman's 2010, *Figures of Invention: A History of Modern Patent Law*,[6] have not only placed intellectual property front and centre in the study of law, science and policy, they have also considerably broadened the scope of scholarship on intellectual property.

To explore further the law and GMO convergence, we have drawn together a range of contributors. Some of them have taken up the challenge of specifically addressing aspects of the intellectual-property-GMO-nexus. These contributions address intellectual property issues of the moment; contemporary twists on perennial themes of patentable subject matter, patents on biological material, scope of biological claims, and so on. For the other contributors, intellectual property has been broadly conceived as a point of contact on the frontiers of newer debates about competition law, disclosure, labelling and information. These eclectic analyses demonstrate the diversity of ways the law and GMOs have become entangled. The stories presented here also demonstrate the complex and exciting evolutionary character of the interactions between law and GMOs.

The various contributions in this volume do not fall neatly into discreet themes. Instead, they reflect something of the complex, messy and sprawling relations that have formed between law and GMOs. We have, accordingly, arranged them alphabetically by author. Without it being our intention, a pattern has emerged none the less. We begin with a pair of chapters in court, or courts, at signal moments of patentability adjudication. Next we have two chapters about GMOs and competition. In the first, competition between the international giants of agricultural GMO, in the second, between tradition and the ability to evolve and innovate. Two further chapters look to the future. The first takes up the gene patent story to look to the fall out of the US Supreme Court's 2013 rulings. The second looks to regulation but with a new twist, regulation of the labelling of GMO food as this regulation moves through higher and higher legislative jurisdictions to the international trade agreements which now might be set to become the final arbiters of GMO risk regulation. Finally we return to court (and the FDA), to two fine grained analyses of the making of GMOs in the mundane procedural workings of law; IP and regulatory. In this travel and return we feel the book mirrors the character of change in this evolutionary relationship. As law and GMOs have grown old

---

5   Biagioli, M., Jaszi, P. and Woodmansee, M. 2011. Making and Unmaking Intellectual Property: Creative Production in Legal and Cultural Perspective. Chicago, Il; University of Chicago Press.

6   Pottage, A. and Sherman, B. 2010. *Figures of Invention: A History of Modern Patent Law*. Oxford, UK: Oxford University Press.

together – their relationship now in its thirties – old themes have found new twists while new issues have succumbed to the structures of old debates.

In the first contribution, '*Cui bono?* Gauging the successes of publicly-funded plant breeding in retrospect', Berris Charnley takes an historical approach to the 2013 US Supreme Court decision in *Bowman v. Monsanto Co*. The case, focused on Monsanto's Roundup Ready resistant GMO-product, exemplifies a particular view of plant breeding. This view attributes successful plant breeding to private corporations and their ability to claim intellectual property. Charnley's historical analysis reveals, however, the widespread historical reliance on public plant breeding to develop agriculture in the US and Britain. What is more, the case reveals the peculiarly extensive nature of intellectual property held over DNA sequences in modern plant gene patents, especially when contrasted to historical constructions of plant breeding work, its aims and products. History reveals that there is nothing inevitable or terminal about the particular juncture at which we find ourselves today. In the recent past public plant breeding was successful on its own terms and without the use of gene patents, contrary to the general view expressed in *Bowman*.

The next entanglement of law and GMOs in the volume is also considered from an historical view, this time, married to the technical detail of European law and the *Directive on the Legal Protection of Biotechnological Inventions*.[7] Stephen Hubicki's '"The story of a love spurned": Monsanto in the United Republic of Soy' provides a detailed analysis of the Court of Justice of the European Union's (CJEU) decision about the European Biotechnology Directive and the intriguing background to this dispute. The dispute centred on whether a patent over a GMO extended to downstream commodity products, in this case soy products (and soy meal) exported from Argentina into Europe. Hubicki's detailed analysis of the tumultuous history of the Directive, aimed at divining the legislative intent embodied in the various provisions of the Directive, is an ambitious project. His analysis, however, brings coherence to a perennial concern in patent law about the importation of products produced abroad using a patented process.

Charles Lawson's 'Competition in the agricultural seeds sector: Patents and competition at a cross-roads?' examines the way intellectual property (essentially patents) and GMOs have converged to raise interesting competition challenges. His analysis shows that even though there are areas of competition conflict these are being resolved by the parties amongst themselves. Many of the potential problems of competition around GMOs have so far been addressed through inter-company deals. Competition and licencing law have, so far, been 'quiet' areas of interaction between law and GMOs, taking place not in the glare of court or legislative negotiations, but rather through the private dealings between the parties in the gaze of competition laws. His conclusion, however, is that there is an imperative in such deals on maintaining existing power among the current GMO producers

---

7   Directive 98/44/EC of the European Parliament and of the Council of 6 July 1998 on the legal protection of biotechnological inventions.

and that this shifts innovative focus in the sector away from new inventions and towards protecting existing inventions and market arrangements. His concern is that the likely consequence of reduced competition is a price squeeze on farmers as they offset higher seed prices – the result of increasingly ordered and non-competitive relations between seed firms – against long-term stability in global commodity market prices.

Also looking to future issues, Karinne Ludlow's 'Regulating for traditional innovation in agricultural organisms' assesses the risk that existing geographic indications (GI) regimes and proposed legal responses to traditional knowledge (TK) will stifle innovation, leading to farmers and communities having to continue using outdated technology and plant materials. She critically assesses the opportunities for protecting traditional innovation in agricultural organisms and what these mean for the adoption of GMOs. Ludlow concludes that if 'traditional' technology – thought of as a counterpoint to modern technology and GMOs – is to be protected then care will need to be taken in creating legal protection that ensures there is space for 'traditional' innovation.

Next Dianne Nicol's '*Myriad Genetics* and the remaining uncertainty for biotechnology inventions' analyses the 2013 US Supreme Court decision in *Association for Molecular Pathology v. Myriad Genetics, Inc* as to whether gene sequences are patentable. She then reflects on the implications of this decision and traces the consequent litigation. She concludes that the impact of the decision may be significant if a broad approach to the interpretation of the decision, and the related decision in *Mayo Collaborative Services v. Prometheus Laboratories, Inc.*, is favoured. Nicol's analysis of recent decisions suggests that both the judiciary and the US patent office appear to be adopting this approach with what could be interesting results for the agricultural biotechnology industry.

Matthew Rimmer's '*Just label it*: Consumer rights, GM food labelling, and international trade' surveys the debate in the US over state, federal and international efforts to engage in GM food labelling. Rimmer's analysis reveals there has been considerable debate about state and federal GM food labelling initiatives and a lack of consensus in the US Congress. The analysis also reveals, however, that the forum for these initiatives is moving to the level of regional and international trade agreements which may stifle domestic initiatives. Rimmer concludes that there may be a need to ensure that consumer rights about food labelling are properly respected and recognised in such regional and international trade negotiations.

Next Jay Sanderson and Fran Humphries' 'Unnaturally natural: Inventing and eating genetically engineered AquAdvantage® Salmon, and the paradox of nature' examines the slippery nature of GMO salmon. Depending on the context in which the GMO salmon are situated, they have been claimed as natural and as something other than natural. This chapter fleshes out the detail of how GMO salmon that are described in a patent claim as something other than natural could also be natural when submitted for FDA safety testing.

The final contribution to the volume, Kieran Tranter's 'Information about information about information: GMOs and law as a "flexible technology"', considers the mundane and routine practice of law in an interlocutory proceeding. In order to work out whom they could sue for infringement, Monsanto needed to find out where its patented GM cotton seeds had gone. In a brief decision, authored in *Monsanto Company v. Syngenta Seeds Pty Ltd*, one of the stars of Australian jurisprudence, Justice Ray Finkelstein weighed up the merits of forcing Syngenta to disclose which of its subsidiaries was in possession of Monsanto's missing seed. Tranter, drawing on Haraway, Latour and Heidigger, argues that what appears to be the mundane and routine practice of law is actually a process of formatting the world. It is through the law, Tranter argues, that GMOs come to exist in the particular way they do in the world.

GMO discourses have overtaken discussions of risk (and the closely related topic of trust) to engage and challenge the frontiers of intellectual property – patents, GIs, TK, subject matter, information, branding – and beyond, into competition law and free speech. Clearly this is just the start of a much broader engagement between GMOs and law (and particularly intellectual property law). As GMO technology becomes increasingly more complex and increasingly more embedded in our lives we can expect to see this area of research blossom. The chapters collected here provide a map to some of the most interesting current issues of GMO and law interactions, mostly centred on intellectual property law. These chapters also point, we hope, to several of the issues most likely to arise in the future as GMO-law discourses spill over into new areas of interaction.

Chapter 2

Cui bono? Gauging the Successes of Publicly-funded Plant Breeding in Retrospect

Berris Charnley

Introduction

In the 2013 US Supreme Court case, *Bowman v. Monsanto Co.*, a particular view of plant breeding and its history emerged, one which attributed successful breeding to plant breeding firms and their ability to claim strong intellectual property.[1] This view reflected the status quo at the time of the case; in the first years of the twenty-first century commercial breeding and strong intellectual property have been in the ascendency. Justice Roberts' court in *Bowman v. Monsanto Co.* ruled in favour of Monsanto and extensive intellectual property protection in plant breeding. In reaching this opinion, the Justices and several contributors considered and dismissed the alternatives: publicly funded plant breeding and weak intellectual property protection. The case documents of *Bowman v. Monsanto Co.* reveal a view of plant breeding history seen from the perspective of a privatised industry and a legal institution that was long-used to dealing with the industry and partly responsible for its stronger intellectual property.

Two features of the pro-commercial, pro-intellectual property view of plant breeding advanced at *Bowman v. Monsanto Co.* are striking. The first is the tendency, apparent in industry briefs presented to the court and in the Justices' remarks during oral arguments, to consider publicly-funded schemes and their aims unimportant, or even counter-productive to agricultural development. The second feature of the case is the Supreme Court's view of what it means to

---

1   An earlier version of this chapter was presented at the Griffith University workshop, 'GMOs Driving Legal Developments – Updates from the Front', Coolangatta, 18 October 2013. I would like to thank the participants at the workshop for several helpful suggestions as to how to develop the paper. Finally, I would like to thank Graham Dutfield, Jay Sanderson, Brad Sherman and Brendan Tobin for providing comments and feedback which have no doubt made this chapter stronger. All remaining errors are, of course, my own.

Bowman v. Monsanto Company Co., 133 S.Ct. 1761 (2013). Case documents relating to *Bowman v. Monsanto Co.* are available at *SCOTUSblog*. [Online]. Available at: http://www.scotusblog.com/case-files/cases/bowman-v-monsanto-co/ [accessed 25 February 2014].

'make' in plant breeding and what this means for infringement of patent claims concerning deoxyribonucleic acid (DNA) sequences embodied in a GMO. The Supreme Court Justices' conceptualisation of making – in this case, the mere appearance of DNA sequences in a plant's genome, however they came to be present, reflects US policy – at the time of the case – of granting extensive intellectual property to commercial plant breeders.

This chapter examines two historical aspects of the court's decision in *Bowman v. Monsanto Co.* On one hand it looks to the historical view of plant breeding revealed in this signal gene-patent case; on the other hand, the chapter analyses what has been called the 'de-historicised gene'; 'discrete, interchangeable, binary ... the path through which it passed from one organism to another, whether in the lifetimes of individuals or across centuries, [makes] no difference to its identity or capacities'.[2] The de-historicised gene, cut adrift from the history of its genesis or propagation, lies at the centre of Monsanto's DNA sequence claims and is the object the court found Bowman to have infringed. However, setting the court documents of *Bowman v. Monsanto Co.* against their historical context, demonstrates in vivid relief that profits are not the only incentive to plant breeding and that a focus on DNA as the locus of creation by contemporary intellectual property law is one that favours profit-making plant breeding at the expense of other schemes which, recognising the fluctuating, historical nature of varieties, rewarded stewardship.

The chapter begins with an outline of *Bowman v. Monsanto Co.* and two controversial pieces of legislation from the 2010s. The context of plant breeding revealed by these legal events is then contrasted with the historical development of agricultural funding and research in the US and Britain. From this historical background we turn back to the *Bowman v. Monsanto Co.* court documents and the view of plant breeding history they reveal. Finally we turn to the narrow conceptualisation of making described in *Bowman v. Monsanto Co.* contrasted with a much broader historical concept of making that was dominant amongst plant breeders in the 1930s.

Outline of *Bowman v. Monsanto Co.*

The most obvious feature of *Bowman v. Monsanto Co.* – which did not turn out to be the test-case Monsanto, Bowman, the anti-GMO lobby or anyone else involved wanted it to be – was just how consistently the court, and several groups filling

---

2   The term 'de-historicised gene' is borrowed from historian-of-science Dominic Berry's work on pure line theory and its failure to influence plant breeding. Berry, D. 2014. The plant breeding industry after pure line theory: Lessons from the National Institute of Agricultural Botany. *Studies in the History and Philosophy of Biological and Biomedical Sciences,* 46, 25–37; 28. See also Palladino, P. 2002. *Plants, Patients and the Historian: (Re)Membering in the Age of Genetic Engineering.* Manchester: Manchester University Press.

amici briefs in support of Monsanto, spoke to a belief that commercial plant breeding is the best means to achieve agricultural development, and intellectual property is the best means to spur commercial plant breeding.

A wide range of groups filed briefs. In total 22 amici briefs were submitted to court. Seventeen briefs were submitted in support of Monsanto, the remaining five in support of Bowman. The American Seed Trade Association (ASTA), American Soybean Association, Pioneer Hi-Bred and Croplife (America and International) filed briefs supporting Monsanto; the Center for Food Safety and Save Our Seeds filed for Bowman. Several more generally intellectual-property-focused groups also filed on both sides; the Intellectual Property Owners Association and New York Intellectual Property Law Association filed for Monsanto and the American Antitrust Institute and Public Patent Foundation filed for Bowman. Finally, several groups from different industrial sectors filed briefs on both sides, for example BSA – The Software Alliance – filed a brief in support of Monsanto while the Automotive Aftermarket Industry Association filed in support of Bowman.

The Missouri-based Monsanto Company has become a focal point for debates around commercial plant breeding, intellectual property and GMOs. One of the anti-GMO movement's concerns is that the application of patent protection to DNA sequences has facilitated a concentration of the seed industry into the hands of a few large corporate entities.[3] Monsanto has certainly not been shy in protecting its intellectual property.[4] However, if the company's attitude to litigation has sometimes verged on the pathological, the philosophy behind such actions is increasingly widespread.[5] Several academics are now promoting the ideas that: (a) commercially-produced GM varieties are the best way to face the

---

3   The Center for Food Safety has been one of the loudest proponents of this view. See Center for Food Safety. 2005. *Monsanto vs. U.S. Farmers.* Washington, DC: Center for Food Safety; Center for Food Safety and Save our Seeds. 2013. *Seed Giants vs. U.S. Farmers.* Washington, DC: Center for Food Safety.

4   One infamous early case was that of Canadian farmer, Percy Schmeiser, sued by Monsanto for planting protected seeds which Schmeiser claimed had blown in on the wind. Interestingly in 2013, this worm turned as organic farmers began suing Monsanto for contamination of their GMO-free crops. See Kondro, W. 2004. Monsanto wins split decision in patent fight over GM crop. *Science,* 304, 1229. See also *Organic Seed Growers and Trade Association, et al., v. Monsanto Company, et al.* S. Ct. 13–303, covered here: Laskawy, T. 2011. Reversing roles, farmers sue Monsanto over GMO seeds. *Grist.* [Online]. Available at: http://grist.org/sustainable-food/2011–03–31-reversing-roles-organic-farmers -sue-monsanto-over-gmo-seeds/ [accessed 24 February 2014].

5   See Kevles, D. *Enforcing intellectual property rights in fruit trees and plants: Contracts, patents and the courts in the 1920s and now,* symposium presentation at 'Intellectual Property and the Biosciences', White Rose IPBio Symposium and Summer School (7–8 July, 2010), University of Leeds. [Online]. Available at: http://ipbio.org/daniel-j-kevles-enforcing-intellectual-property-rights-in-fruit-trees-and-plants-contracts-patents-and-the-courts-in-the-1920s-and-now.htm [accessed 24 February 2014].

challenges of population growth and climate change and (b) strong intellectual property encourages new GM varieties to market.[6]

Bowman v. Monsanto Co. stands at the end of a series of cases which Monsanto has brought against farmers it believed to be infringing on its patent protected Roundup-Ready soybeans.[7] A complicating feature of these cases has been Monsanto's technology agreement contract. A document signed with any purchase of Monsanto's seed – the contract which requires farmers to comply with Monsanto's restrictions on saving and replanting of seeds (among other things) – means these cases have been about the interplay between contract law and intellectual property law.[8]

For Indiana soybean farmers there are one and a half growing seasons in the year. The year's first crop is the one on which they stake their livelihood. But with luck it is possible to squeeze a second, riskier, crop into the year. Vernon Hugh Bowman purchased Monsanto's soybeans for his first crop of the year and signed the company's technology agreement with these purchases. However, for his second crop, Bowman did not want to spend money on seeds that might fail if the weather was bad. To get around this problem, in 1999, Bowman – while continuing to purchase seeds from Monsanto for his first crop – began purchasing seeds for his second crop from a grain elevator. Grain elevators are large storage facilities where soybeans are collected into a generic pool to be channelled into the livestock and human feed industry. Bowman bought beans from the elevator at a much lower price than he would have paid Monsanto. Guessing that within this mix of pooled beans there would be some descendants of Monsanto's patented Roundup-Ready plants, Bowman planted his elevator purchases. When the beans grew he treated them with Monsanto's weed killer – Roundup – and through this selection established which plants were, like Monsanto's, Round-up resistant. Bowman harvested beans from these plants to use for his yearly second crop.[9]

---

6 See for example, Tait, J. and Barker, G. 2011. Global food security and the governance of modern biotechnologies. *EMBO Reports*, 12, 763–768; Thompson, P. 2011. *Agro-technology: A philosophical introduction*. Cambridge: Cambridge University Press on the necessity of GMOs and intellectual property over GMOs. See also, Marie Brown, N. and Fedoroff, N. 2004. *Mendel in the kitchen: A scientist's view of genetically modified foods*. Washington: Joseph Henry Press and Kingsbury, N. 2011. *Hybrid: The history and science of plant breeding*. Chicago: University of Chicago Press.

7 See *Monsanto Company v. David,* 516 F.3d 1009 (2008); *Monsanto Company v. Parr, 545* F.Supp.2d 836 (2008); *Monsanto Company v. Vanderhoof* 2007 WL 1240258 (2007); *Monsanto Company v. Strickland,* 2007 WL 3046700 (2007); *Monsanto Company v. Scruggs, 459* F.3d 1328 (2006); *Monsanto Company v. Good,* 2004 WL 1664013 (2003); *Monsanto Company v. McFarling,* 302 F.3d 1291, (2002). *Monsanto Company v. Trantham, 156* F.Supp.2d (2001) and *Monsanto Company v. Dawson,* 2000 WL *33953542* (2000).

8 For a roundup of previous cases and on the interplay between intellectual property and contract law, see Lawson, C. 2011. Juridifying the self-replicating to commodify the biological nature future: patents, contracts and seeds. *Griffith Law Review*, 20(4), 851–882.

9 For an outline of the case see Crouch D. 2013. US Government brief: Farmer who purchases commodity soybeans cannot replant those beans without committing patent

Bowman believed he had found a way to get around Monsanto's technology agreement (which he had not signed when making purchases from the grain elevator) and the company's patent claims. On Bowman's reading the exhaustion doctrine curtailed Monsanto's patent rights after the first sale of seeds. In brief, the exhaustion doctrine – established in the nineteenth century – safeguards the right freely to use purchased items (including reselling them) without deference to a patent holder's rights. The original intention behind the doctrine was to support the assumption that a purchase entails a whole transfer of property from vendor to customer. One thing users are not allowed to do – in most circumstances – is use their purchased item to make copies.[10] Accordingly, one of the key battle grounds on which the exhaustion doctrine has evolved has been over the point at which use and (re)making can be distinguished. Maintenance and repair work are the shades of grey which have blurred that distinction.[11] Relying on this interpretation of the doctrine, Bowman believed he was buying seeds from which Monsanto's intellectual property had been exhausted.

In 2007 Monsanto arrived in Indiana to investigate Bowman's crops. When they discovered that Bowman's second crop was resistant to Roundup, Monsanto sued for infringement of its US Patent Nos. 5,352,605 and RE 39,247E (which included claims covering DNA sequences relating to resistance). The Federal Court of Indiana found in Monsanto's favour: Bowman's activity constituted infringement. Damages of $84,000 were awarded to Monsanto. When Bowman appealed to the United States Court of Appeals for the Federal Circuit, the lower court's judgement was upheld. In late 2011, Bowman filed petition for a writ of certiorari with the Supreme Court on the grounds that:

> Patent exhaustion delimits rights of patent holders by eliminating the right to control or prohibit use of the invention after an authorized sale. In this case, the Federal Circuit refused to find exhaustion where a farmer used seeds purchased in an authorized sale for their natural and foreseeable purpose – namely, for planting. The question presented is: Whether the Federal Circuit erred by (1) refusing to find patent exhaustion in patented seeds even after

---

infringement. *Patently O.* [Online]. Available at: http://patentlyo.com/patent/2013/01/us-government-brief-farmer-who-purchases-commodity-soybeans-cannot-replant-those-beans-without-committing-patent-infringemen.html [accessed 24 February 2014]; Liptak, A. Supreme Court supports Monsanto in seed-replication case. *The New York Times.* 14 May 2013, B3. [Online]. Available at: http://www.nytimes.com/2013/05/14/business/monsanto-victorious-in-genetic-seed-case.html?_r=0 [accessed 14 May 2014].

10   However, Art. 30 of TRIPS provides some space for legal making under certain conditions.

11   See the extended discussion of exhaustion in Lawson above n 8, 861–873; Sheff, J. 2013. Self-replicating technologies. *Stanford Technology Law Review,* 16(2), 229–256 and Ghosh, S. 2013. *The implementation of exhaustion policies: Lessons from national experiences*; ICTSD Programme on Innovation, Technology and Intellectual Property; Issue Paper No. 40. Geneva: International Centre for Trade and Sustainable Development.

an authorized sale and by (2) creating an exception to the doctrine of patent exhaustion for self-replicating technologies?[12]

In October 2012 the petition was granted. Oral arguments were held in February 2013 and on 13 May 2013 Justice Kagan delivered the court's opinion. In a unanimous, yet narrowly-tailored, decision the court held that Bowman's activity was not protected by the exhaustion doctrine and upheld the lower courts' decisions and award of damages.

Commercial Plant Breeding in the Twenty-first Century

The court's opinion refused to define any sort of precedent but instead adopted a 'business as usual' statutory interpretation. However, if the case was not precedent setting, the status quo it revealed was a positive attitude to commercial breeding and intellectual property protection, embodied in many of the briefs filed in support of Monsanto. The section headings of the ASTA's amici brief make such an ethos particularly obvious. The brief begins with Section I, the section title proclaiming the benefits, but also the financial costs, of patented seed technology:

> Section I. Patented Seed Technology provides numerous benefits to society but is costly and time consuming to research and develop.[13]

The brief runs through to Section III, whose title makes obvious that intellectual property is an essential incentive, before raising the spectre of how the public might be deprived, if commercial plant breeding could not continue:

> Section III. Removing protection for each generation of patented seed would devastate the nation's seed industry, evaporate investment in patented seed technology and deprive the public of this technology's current and future benefits.[14]

In the statement of interest contained in their brief ASTA explicitly laid out their members' position, tying together the financial costs, putative benefits and need for intellectual property:

> ASTA members annually invest billions of dollars researching and developing patented seed technology to make American agriculture more productive and the

---

12   *Bowman*, above n 1, writ of certiorari, question presented, i.

13   *Bowman*, above n 1, Brief Amicus Curiae of the American Seed Trade Association in support of respondents, i–ii.

14   Ibid.

> Nation's food supply more plentiful and nutritious. To protect this investment
> these entities seek patent protection for their discoveries.[15]

The view of agricultural development and plant breeding advanced by ASTA – as an activity best conducted by private firms incentivised by intellectual property – is based on the incentive argument: the view that intellectual property operates to incentivise innovation, through the promise that research expenditure can be recouped during a temporary period of monopoly. However, intellectual property is not the only way to create a monopoly. The biological features of some plant species and regulatory systems for plant varieties can also restrict the sale of seeds into the hands of a few companies.

In the US, from the 1930s onwards, commercial plant breeding companies came to dominate plant breeding and the seed market. This capture was largely facilitated by double cross F1 hybrid corn, a new type of agricultural variety that included its own biological means of dissuading farmers from saving seed or breeding plants themselves and encouraging them to return each year to the seed merchant for fresh stocks.[16] The relative un-importance of corn in much of European agriculture goes some way to explaining the later ascendance of commercial plant breeding outside of the US. The most important crop in Europe and Britain – wheat – offers no such biological control of seeds despite numerous attempts to produce double cross F1 hybrid wheat varieties. Unable to repeat this monopolising trick with all plant species, commercial breeders have increasingly turned to intellectual property as a means of securing control over seed supplies. We return to the type of intellectual property that has resulted in the last section of this chapter. However, a second area of the law, regulation, has also aided commercial plant breeding in recent years.

Two pieces of legislation, one from the US and one from the EU, both on their respective legislature's agendas in 2013, illustrate the dominance of commercial plant breeding and the concentration of seed control. In the US a rider on the *Consolidated and Further Continuing Appropriations Act 2013*, known as the Farmer Assurance Provision, was momentarily sworn into legislation.[17]

---

15   Ibid., 2.

16   The classic work on the problems of double cross hybridisation as a method of plant breeding is from rural sociologist Jack Kloppenberg and Marxist scholars Richard Lewontin and Richard Levins: see Kloppenburg, J. 1988. *First the seed: The political economy of plant biotechnology*. Cambridge: Cambridge University Press; Lewontin, R. 1990. The political economy of agricultural research: The case of hybrid corn, in *Agroecology* edited by C. Carroll, J. Vandermeer and P. Rosset. New York: McGraw Hill, 613–626; Lewontin, R. and Berlan, J.-P. 1986. Breeders' rights and patenting life forms. *Nature*, 322(6082), 785–788; Levins R. 1986. Science and progress: Seven developmentalist myths in agriculture. *Monthly Review*, 38(3), 13–20.

17   Farmer Assurance Provision, Section 735 (formerly Section 733) of US H.R. 933 – *Consolidated and Further Continuing Appropriations Act 2013*, 113th Congress (2013–2014). See the chapter from Matthew Rimmer in this volume for more on the

The provision was renamed the Monsanto Protection Act by opponents who claimed the provision exempted GMO varieties from judicial challenges over their safety. On this view, the bill meant that the US Department of Agriculture (USDA) could overturn court rulings on the safety of GMO varieties and allow their planting. Enactment of the provision caused a backlash, including a *March against Monsanto* and the wording was stripped from the wider act.[18] The removal of this provision has generated much hope in the food safety movement that Monsanto's influence in Washington is waning. However, the fact that it was enacted speaks just as loudly to the contemporary situation as its overturning does to the future. Monsanto's operational requirements – that safety certifications on new varieties need to be stable to be profitable – were ranged directly against the public's right to challenge safety certifications in court. For a moment, at least while nobody seemed to be watching, the US administration explicitly prioritised Monsanto's needs over considerations of public access, through courts, to the regulatory system.

In Europe new seed regulations, now being drafted, would make it necessary for plant breeders to register all new varieties. Part of the registration requirements would entail something like distinctness, uniformity and stability testing and productivity testing.[19] These regulations would, furthermore, restrict farmers to using only seed that have been tested and registered in this manner.[20] The effects of the proposed new regulations would be somewhat similar to the results of tight laboratory safety procedures and strong regulation in the biotech industry. The cost

---

provision and its impact on GMO food labelling. See also Godoy, M. 2013. Did Congress just give GMOs a free pass in the courts? *NPR: The Salt,* [Online]. Available at: http://www. npr.org/blogs/thesalt/2013/03/21/174973235/did-congress-just-give-gmos-a-free-pass-in-the-courts [accessed 24 February 2014].

18   See the *March against Monsanto*'s official website. [Online]. Available at: http:// www.march-against-monsanto.com/ [accessed 24 February 2014] and Alliance for Natural Health, 2013. Success! The Monsanto Act has been repealed-this time for good! [Online]. Available at: http://www.anh-usa.org/monsanto-protection-act-has-been-repealed/ [accessed 24 February 2014].

19   European Commission. 2013. On the production and making available on the market of plant reproductive material. Brussels, 6.5.2013 COM(2013) 262 final. 2013/0137 (COD).

20   On similar regulations in the French national context see Anvar, S. Introduction to 'plant variety product licensing' and its impact on plant breeding and plant intellectual property rights (PIPR), in *Living Properties: Making Knowledge and Controlling Ownership in Modern Biology,* J.-P. Gaudillière, D.J. Kevles and H.-J. Rheinberger (eds)., Berlin: Max Planck Institute for the History of Science Preprint 382, 57–64. On seed systems more generally see: Louwaars, N. 2007. Seeds of confusion: The impact of policies on seed systems. Unpublished PhD dissertation, Wageningen University. Formal seed systems, as proposed in this EU legislation, can be fruitfully contrasted to participatory seed breeding increasingly conducted by community groups in developing countries. See Tapia, M. and Tobin, B. 2013. Guardians of the seed: the role of Andean farmers in the caring and sharing of agrobioversity, in *Common pools of genetic resources: Equity and innovation in international biodiversity law,* edited by E. Kamau and G. Winter. Abingdon: Routledge, 79–100.

of compliance, so the industry claims, acts as a bar to conducting research for companies without substantial capital. In this case the new legislation presumes that the only reason for plant breeding is for profit: hence the regime's close alignment with the criteria already established by the International Union for the Protection of New Varieties of Plants (UPOV) for plant breeder's rights. Plant breeders working for their own bliss are not only invisible in this legislation; they are actively excluded from the production and sharing of new varieties by the costs of registration. In response to protests from small and organic plant breeders and farmers, a modified version of the legislation has been drafted. However, even in this new form, the proposed regulatory scheme will probably act as a bar to less capitalised plant breeders operating in the EU seed market.[21]

The rise of commercial breeding has not been entirely welcomed and concentration of breeding into the hands of an increasingly small number of corporations has caused much concern. These problems were recently acknowledged in the British Government's Foresight report on the future of agriculture:

> Over the last two to three decades a relatively small number of companies have come to dominate in the global food supply chain. This trend is apparent all along the supply chain, from agri-business (including seeds, crop protection) through to commodity wholesalers, manufacturers and retailers. Concerns have been raised regarding the exercise of this concentration of corporate power, for example in retail markets and purchase contracts with suppliers (particularly smaller farmers); wider public access to agricultural intellectual property and the transparency of governance in the food system.[22]

It is important to note that although commercial entities now dominate the production and sale of seeds, a great deal of publically funded research is still being conducted. However, as a result of an ongoing and often ideological rearrangement of the sector, publicly funded research is largely aimed at basic research, the results of which are then passed to commercial breeders to prepare for market and, often, protect with a plant variety registration or patent. This is the near and far market distinction that was promoted so vehemently in late-Thatcherite economics and is

---

21   For analysis of the new regulations see Rabesandratana, T. 2013. Overhaul of EU seed regulations triggers protests. *Science Insider*. [Online]. Available at: http://news.sciencemag.org/europe/2013/05/overhaul-e.u.-seed-regulations-triggers-protests [accessed 24 February 2014]. The campaign is being taken up by several groups, see for instance, EU Seed Legislation Reform [Online]. Available at: http://www.saveourseeds.org/en/dossiers/eu-seed-regulation.html [accessed 24 February 2014].

22   Government Office for Science. 2011. *Foresight. The future of food and farming executive summary*. London, UK: The Government Office for Science, 23. Better access to intellectual property and to intellectual property protection were the major suggestions, short of direct market intervention, from this group as to how to tackle such industry concentration.

being revived in the twenty-first century, which underpins the policy decision that activities near to the market (such as commercialisation) should be conducted by the private sector, while those far from the market (such as basic research) should be conducted by the public sector.[23]

The foregoing discussion illustrates the relative dominance of seed production and sales in the early twenty-first century by commercial entities which are favoured by contemporary legislation.[24] That such entities should be chiefly concerned with profit making, ('greater yields, healthier crops and higher revenues than ever') and that the rules of the game prioritise this motive and facilitate its prosecution will be no surprise.[25] However, the alliance of groups protesting against seed legislation in the US and EU, reveals a quite different picture of non-commercial plant breeding. In the US, public protest was raised against the consolidation of corporate plant breeding power; in the EU, protest by amateur and organic plant breeders was instead directed against the exclusion of non-commercially motivated breeders from the market.

The history of publicly funded plant breeding in the nineteenth and twentieth centuries provides another view of how and why plant breeders might be motivated to produce seeds without profits. In the nineteenth century US Republic and the British Empire of the late nineteenth century plant breeding was deployed for the 'good of the nation' and 'the empire' in ways that demanded that the virtuous would give freely of their time.

Historical Context of Agricultural Research

In the early twentieth century publicly funded plant breeding was enjoying a heyday in the US and across much of Europe.[26] Beginning in the 1830s the United States Patent and Trademark Office (USPTO) collected and then redistributed

---

23   On this policy and the privatisation of the UK's Plant Breeding Institute see Webster, A. 1989. Privatisation of public sector research: the case of a plant breeding institute. *Science and Public Policy*, 16 (4), 224–232. More generally, see Edgerton, D. and Hughes, K. 1989. The poverty of science: A critical analysis of scientific and industrial policy under Mrs Thatcher. *Public Administration*, 67(4), 419–433 on Thatcherite innovation policy.

24   For more evidence of the sway held by such corporations over the US legislature see the chapter on GM labelling from Matthew Rimmer in this volume.

25   ASTA and Croplife. *The guide to seed treatment stewardship*. [Online]. Available at: http://seed-treatment-guide.com/about/overview/ [accessed 5 May 2014].

26   For an overview of that history with reference to private and public research themes see Pardey, P. and Beintema, N. 2001. Slow magic: Agricultural R&D a century after Mendel. Washington: IFPRI. See Harwood, J. 2012. *Europe's Green Revolution and others since*. London, Routledge (on the constellation of public funding research institutes in Germany and across central Europe and their historical development). See Palladino, above n. 2 and Charnley, B. 2012. *Agricultural science, plant breeding and the emergence of a Mendelian system in Britain, 1880–1930*. Unpublished PhD Thesis. University of Leeds (on the development of publicly funded research in Britain and its colonies).

increasingly large quantities of foreign seed.[27] Collection was orchestrated through a network which became increasingly large and formalised over time. Initially, naval officers and embassy staff operated as collecting agents. This work was conducted in free time and often without adequate funding to ensure the safe passage of samples back to the US. During this early period the USPTO was involved in an oversight role without providing any large amount of direct funding. This was work – essential for plant breeding – undertaken for the good of the nation. If collection was run on a shoe string in these early years, so was distribution. Parcels of seeds were sent out under congressional franks (a free postal privilege of congressional members). Through the rest of the nineteenth century the scheme was funded, defunded and refunded according to political whim but, in one form or another, it persisted until 1924. In later years specialist collecting agents were hired and several seed houses in Britain, France and across Europe were enlisted as official suppliers. By the 1880s the programme was being run by the newly formed USDA (that was initially a sub-department of the USPTO) and roughly 10 million parcels of seed were being distributed to farmers annually.[28]

Whether the scheme operated to provide commercial quantities of seed to farmers or samples for trialling new varieties in new geographical regions remains an open historical question. However, it is clear that one of the main aims of the scheme (however it sought to prosecute this aim) was to expand US agricultural acreage. Undoubtedly, seed provided by the scheme, along with seeds brought from Europe by immigrants, were an essential feature of the westwards expansion of the wheat belt, the equivalent expansion of maize acreage and the horticultural colonisation of California and Florida.[29] Furthermore, during this dramatic colonisation of new growing areas, often with less favourable conditions, or with new pests and diseases, yields remained somewhat constant. Biological innovation, including seed distributed by the US government played an important maintenance role in protecting yields against adverse conditions.

---

27 On the scheme see: Kloppenburg, above n 16; Blair, D. 1999. Intellectual property protection and its impact on the US seed industry. *Drake Journal of Agricultural Law*, 4(1), 297–331; Fowler, K. 2000. The Plant Patent Act of 1930: A sociological history of its creation. *Journal of the Patent and Trademark Office Society*, 82(9), 621–44; Klose, N. 1950. *America's crop heritage: The history of foreign plant introduction by the Federal Government.* Ames: Iowa State College Press.

28 Fowler, ibid., 623.

29 See Olmstead, A. and Rhode, P. 2008. *Creating abundance: Biological innovation and American agricultural development.* Cambridge: Cambridge University Press. A similar agricultural expansion, also assisted by public plant breeding, was occurring in several places around the world at the turn of the century, including Australia, see Olmstead, A. and Rhode, P. 2007. Biological Globalization: The Other Grain Invasion, in *The New Comparative Economic History*, edited by Timothy J. Hatton, Kevin H. O'Rourke and Alan M. Taylor. Cambridge, Mass.: Massachusetts Institute of Technology Press, 115–140.

Several pieces of US legislation, enacted concurrently with the free seed scheme, established even more direct government involvement in agriculture. The *Morrill Act 1862* established the land-grant universities, a series of institutions that received land (and later funds) from government grants 'in order to promote the liberal and practical education of the industrial classes in the several pursuits and professions in life', agriculture included.[30] The *Hatch Act 1887* established funding of US$15,000 for each of the land-grant universities to found experimental stations, which became hotbeds of progressive plant breeding.[31] It was at these institutions that genetics found its most enthusiastic reception in the US.[32] Finally, the *Smith-Lever Act 1914* extended this funding and gave experimental stations responsibility for a new cooperative extension programme intended to educate farmers in the new scientific farming.[33]

The free seed scheme and public plant breeding initiatives of the US government facilitated a huge expansion of agriculturally productive land and the maintenance of yields in these new areas. In the period 1850–1940 the amount of land in farms grew from around half a million square miles to over 1.75 million square miles. Total output grew steadily, if unspectacularly, over the same period.[34] As well as expanding acreage, the F1 double cross hybrid maize varieties which so aided the development of commercial plant breeding in the US, were a product of publicly funded plant breeding.[35]

In Britain, government intervention in agriculture has a similarly thoroughgoing history, albeit one that starts slightly later in the century.[36] In 1890 the Residue Grant – known as the whiskey money – rolled into government.[37] These were funds raised through taxes on alcoholic beverages. The idea had been to redistribute these extra funds to the holders of alcohol vending licences. However, in a context in which the temperance movement still held a great deal of influence, this idea

---

30   *Morrill Act 1862* (7 U.S.C. § 301 *et seq.*).

31   *Hatch Act 1887* (ch. 314, 24 Stat. 440, enacted 1887–03–02, 7 U.S.C. § 361a *et seq.*).

32   On the reception of Mendel's work in the US see Kimmelman, B. and Paul, D. 1988. Mendel in America: Theory and practice, 1900–1919, in *The American Development of Biology* edited by R. Rainger, K. Benson and J. Maienschein. Philadelphia: University of Pennsylvania Press, 281–310.

33   *Smith-Lever Act 1914* (ch. 79, 38 Stat. 372, enacted 1914–05–08, 7 U.S.C. § 341).

34   Alston, J., Andersen, M., James, J. and Pardey, P. 2010. *Persistence Pays U.S. Agricultural Productivity Growth and the Benefits from Public R&D Spending.* London: Springer, 15–17.

35   On the history of hybrid maize see Kloppenburg, Lewontin and Levins above n. 16 and Crow, J. 1998. 90 Years Ago: The Beginning of Hybrid Maize, *Genetics*, 148, 923–28.

36   For the general contours of this history see Russell, E. 1966. *A History of Agricultural Science in Great Britain, 1620–1954.* London: Allen and Unwin. esp. chapter 8 and Palladino above n 2.

37   Richards, S. 1988. The South-Eastern Agricultural College and public support for technical education, 1894–1914. *The Agricultural History Review,* 36(2), 172–87.

was abandoned. The newly available funds were instead diverted toward the county councils to provide technical education. The South Eastern Agricultural College at Wye was one of the first institutions to receive funds. As the century turned several university agricultural departments and more technical institutions such the College of Agriculture and Horticulture at Holmes Chapel in Cheshire and the Midland Agricultural and Dairy Institute at Kingston, Nottinghamshire were receiving whiskey money.[38]

Previous government interventions in industry had focused on ad hoc provisions which allowed certain privileges to a specific group. For instance in the saltpetre industry, government granted saltpetremen access to private land to collect saltpetre – a key weapon in early colonial expansion.[39] In contrast, schemes funded with whiskey money where the result of direct government intervention in industry development through the funding of technical instruction and research. The government's Board of Agriculture was renamed and expanded several times in the next quarter century to reflect this new direct role in industry.[40]

In 1909 David Lloyd George announced in his 'People's Budget' the creation of a £1,000,000 fund for rural development, including funding for plant breeding. In the following years, and especially after the First World War, this fund grew significantly.[41] In 1912, a Plant Breeding Institute was established at Cambridge using these funds. After the war, the National Institute of Agricultural Botany (NIAB) was established with 50 per cent of its funding coming from government funds. Increased post-war reconstruction funding, combined with legislative changes which created the *Seed Testing Act 1920* – replete with enforcement mechanisms and funding for an Official Seed Testing Station (a pre-war institute rehoused in NIAB's new buildings after the war) – amounted to a quasi-nationalised system of plant breeding, prosecuting what it claimed to be the latest genetic methods of breeding. In a context dominated by publicly bred varieties, commercial plant breeding firms struggled to stay in business.[42] In the nineteenth

---

38   Ibid., 185. See also Thomas, W. 2012. Agricultural colleges in Britain. *Ether Wave Propaganda.* [Online]. Available at: http://etherwave.wordpress.com/2012/03/26/agricultural-colleges-in-britain/ [accessed 25 February 2014].

39   On the case of saltpetre see Cressy, D. 2013. *Saltpeter: The mother of gunpowder.* Oxford: Oxford University Press.

40   The Board of Agriculture, which had been re-established by the Board of Agriculture Act 1889 was renamed the Board of Agriculture and Fisheries in 1903 and, from 1919, the Ministry of Agriculture and Fisheries. During the war a Department of Scientific and Industrial Research was also established, see Clarke, S. 2010. Pure Science with a Practical Aim: The Meanings of Fundamental Research in Britain, circa 1916–1950, *Isis,* 101, 285–311.

41   See Olby R. 1991. 'Social imperialism and state support for agricultural research in Edwardian Britain', *Annals of Science,* 48, 509–26 and Brassley, P. 1995. 'Agricultural research in Britain, 1850–1914: Failure, success and development', *Annals of Science,* 52, 465–80.

42   On fraught relations between public and commercial plant breeders in Britain see Charnley, B. 2013. Why didn't an equivalent to the US Plant Patent Act of 1930 emerge in Britain? Historicising the boundaries of un-patentable innovation, in *The Intellectual*

century a majority of the plants grown on British farms came from saved seed.[43] By the early 1980s, 90 per cent of Britain's wheat seeds were derived from public funding.[44] When the Plant Breeding Institute was privatised in 1987, Unilever purchased the near-to-market aspects of the Institute – which was broken up in the process – for between $100 and $130 million.[45]

In Britain, much of the motivation for this activity stemmed from a desire to increase home production and manage agriculture in the tropical colonies more efficiently.[46] This was a process from which small scale farmers, in Britain and abroad, were often excluded. In the US case 'agricultural development' was also a process that had winners and losers. Well capitalised farmers were set to reap the benefits of scientifically aided agriculture, developed with public funding, in ways that were unavailable to their poorer rivals. These sorts of considerations militate against any simplistic reading of publicly funded agricultural research as being as good as, or better than, commercially funded plant breeding. However the charge laid against historical publicly funded plant breeding at *Bowman v. Monsanto Co.* was that these schemes failed to develop agriculture.

The History of Plant Breeding Presented in the *Bowman V. Monsanto Co.* Court Documents

During Mark Walters's oral arguments for Bowman, Justice Roberts interrupted Walters in mid-flow and demanded, 'Why in the world would anybody spend any money to try to improve the seed if as soon as they sold the first one anybody could grow more and have as many of those seeds as they want?'[47] An even more explicit view of plant breeding history was included in the court documents. In the amici brief put forward in support of Monsanto by various US soybean growers' associations, collectively the American Soybean Association, the Association

---

Property and Food Project: From feeding the world to rewarding innovation and creation edited by C. Lawson and J. Sanderson. Farnham: Ashgate, 103–23.

43   Brassley, P. 2000. Crop Varieties, in *The Agrarian History of England and Wales, 1850–1914*, edited by E.J.T. Collins. Cambridge, Cambridge University Press, 522–32.

44   Thirtle, C. *et al.* 1998. The rise and fall of public sector plant breeding in the United Kingdom: A causal chain model of basic and applied research and diffusion, *Agricultural Economics*, 19, 127–43, 141.

45   See Palladino, above n 2, 39; Stewart, A. 1987. Sale of lab to Unilever endorsed. *The Scientist*, [Online] Available at: http://www.the-scientist.com/?articles.view/article No/8971/title/Sale-of-Lab-To-Unilever-Endorsed/ [accessed 25 February 2014] and Murphy, D. 2007. *Plant Breeding and Biotechnology: Societal context and the future of agriculture*. Cambridge: Cambridge University Press, 132.

46   On British plans for more efficient, scientific, utilisation of tropical resources see Hodge, J. 2007. *Triumph of the Expert: Agrarian doctrines of development and the legacies of British colonialism*. Athens: Ohio University Press.

47   *Bowman*, above n 1, Oral arguments, 3, ln. 21–4.

claimed that the US free seed scheme actually stifled innovation in the seed industry. The US Government, on this view, acted as an incumbent in the seed market, dissuading investment in research by seed companies. The evidence brought forward for this claim is a putative drop of five bushels per acre in the soybean crop yield from 1866–1930.[48] This statistic hides the massive expansion of the area under agriculture (often into less agriculturally favourable areas) that occurred during this period. On one tally the soybean crop went from less than 50,000 acres in 1907 to around 5.5 million in 1935.[49] Much of this expansion (even in the years after the end of free seeds in 1924) was the result of varietal tinkering, initially facilitated by free seed programmes and later by research at the land grant universities.

The view of historical plant breeding typified in the Association's brief delineates the possibilities for future research by narrowing the range of legitimate aims to which research can be put. Having judged a particular scheme a failure, why in the world would anyone want to replicate that? The Association's mistake in taking yield to be the only measure of a successful agricultural sector is part of a wider history of trait selection which also has ramifications for the current GMO debate. Improving varieties can mean a huge range of things from environmental adaptation, increasing disease or insect resistance, through packaging and storage qualities to experiential qualities at the point of consumption, including but not limited to taste.[50] One of the most consistently highlighted problems with first and second generation GMOs is how little they improved the consumers' experience. If improving quality (not just for farmers but also for consumers) was the Association's aim then on this account their efforts have been a failure.

Bringing the history of publicly funded plant breeding back into view makes two points obvious. Firstly, incentives to innovation have not always been financial. Personal passion, considerations of the good of the nation and national agricultural development have been just as effective in stimulating individuals and institutions in the past. Secondly, this history reveals the widely different aims of commercial and publicly funded plant breeding. The first of these points undermines the incentive to innovate argument – that innovators need strong intellectual property to incentivise them. A second line of argument behind strong intellectual property is the substitution argument – the idea that absent strong rights to exclusivity, purchasers of seed can compete directly with vendors in the market.[51] On the Supreme Court's narrow construction of making, it might seem as though Bowman could have posed a threat to Monsanto by competing with them. However, on a broader construction of making this argument breaks

---

48   *Bowman v. Monsanto Co.* S.Ct. 11–796, Brief of American Soybean Association Illinois … as amici curiae in support of respondents, 10–13.

49   Olmstead and Rhode 2008, above n 29, 278.

50   See Tait and Barker, above n 6, Figure 1, 765 for a glimpse of the changes to trait focus now underway.

51   See Sheff, above n 11, on the substitution argument.

down. Indeed, Bowman continued to buy Monsanto's seeds for his first crop because there were parts of Monsanto's breeding process which he did not have the time or resources to replicate.

The Supreme Court's Concept of Making

The Supreme Court's decision in *Bowman v. Monsanto Co.* was blunt: 'The question in this case is whether a farmer who buys patented seeds may reproduce them through planting and harvesting without the patent holder's permission. We hold that he may not.'[52] A key part of Bowman's defence was the 'blame the bean' argument: Bowman did not make Monsanto's Roundup-Ready beans, having planted his grain elevator purchases (which he claimed as a legitimate use) the beans took care of the creative process. The court rejected this argument, however they failed to create a clear picture of what Bowman actually had done by way of making. On several occasions, as in the quote above, the court referred to making as constituted by the acts of 'planting and harvesting'. This phrase appeared front and centre at the top of the opinion. The application of Roundup (to select for Roundup-resistant seeds) was mentioned as part of Bowman's activities, but as of secondary importance. Yet if Bowman had simply planted and harvested he would hardly have reproduced Monsanto's seeds (note the plural used in the opinion above). Before Bowman treated his elevator purchases with Roundup he was in a similar position to farmers with windblown contamination of GMO seeds. Although Monsanto have promised not to sue in such cases, *Bowman v. Monsanto Co.* has upheld their right to do so in the future. While this possibility seemed to concern Justice Kagan during oral arguments, when she noted, addressing Monsanto's lawyer, Seth Waxman, 'your position ... has the capacity to make infringers out of everybody ... seeds can be blown onto a farmer's farm by wind ... and the ... farmer is infringing', these concerns disappeared from the Court's opinion.[53]

Rather than analysing the precise biological elements of making and, crucially, the relation of Monsanto's protected DNA sequences to the plant as a whole, there was much talk by Justice Breyer of 'magic boxes' and intertwined bunches of grass.[54] These arguments were intended to show that the method of making was irrelevant. During Walters' oral arguments, as a preamble to his magic box thought experiment, Justice Breyer made a telling remark. The case involved several generations of seeds and plants. Breyer interrupted Walters to crack-wise, 'There are three generations of seeds. Maybe three generations of seeds is enough.'[55] Breyer's joke refers to one of the most reviled instances of

---

52   *Bowman*, above n 1, 1765.
53   *Bowman*, above n 1, Oral arguments, 41, ln. 7–14.
54   Ibid. 8, ln. 2–15.
55   Ibid. 7, ln. 8–9.

value judgement in American judicial history (he was also wrong. Bowman had planted, treated and grown several generations in the years before Monsanto arrived in Indiana).

In *Carrie Buck v. John Hendren Bell*, a case which also focused on reproduction, the issue at question was the State's right to forcibly sterilise the daughter of a family who had been deemed 'imbeciles': a pseudo-scientific term at the time.[56] The Supreme Court's remit in hearing the case was to adjudicate on whether Virginia's sterilisation laws contradicted fourteenth amendment guarantees to due process. Justice Oliver Wendell Holmes, Jr. in concluding his argument, declared that: 'Three generations of imbeciles are enough.'[57] In reaching its decision, to sanction forced sterilisation, the court claimed to be acting on the latest scientific evidence. Indeed, Virginia's sterilisation law was derived from model legislation drafted by Harry Laughlin of the Eugenics Record Office based at the Cold Spring Harbor Laboratory. Judging scientific standards retrospectively is poor sport, however, and the science on which Holmes relied – as to judging what constituted imbecility and the efficacy of sterilisation – was far from universally accepted in 1927.[58] The case is now remembered as an example of the use of science to dress up ideological views as objective. Along with Justice Scalia's abstention in part from the Supreme Court's decision in 2013's other stand-out gene patent case – *Association for Molecular Pathology v. Myriad Genetics, Inc.* – on the grounds that the science was beyond him, Breyer's remark indicates that the court has a less advanced understanding of the science in these cases than it sometimes suggests.[59] Remembering Holmes's opinion in *Buck v. Bell* is a fair warning against trusting the scientific basis of the Court's opinion.

The products of plant breeding – what it is that plant breeders suppose they produce – are a moving historical target. Given this ambiguity about what plant breeders take themselves to be making, the court's hard and clean definition looks difficult to sustain. In the 1950s and 60s, despite the US *Plant Patent Act 1930* and UPOV, and the legal arguments around inventorship behind such legislation, many plant breeders still saw themselves as stewards of somewhat natural variability, rather than as inventors. The varieties they produced had to be constantly maintained, otherwise a variety might 'run out', becoming heterogeneous and unruly. This stewardship role – enshrined in breeders' practices – was a key selling point in catalogues which proclaimed the length of time varieties had been maintained and purified. In accordance with population genetics of the period,

---

56 *Carrie Buck v. John Hendren Bell, Superintendent of State Colony for Epileptics and Feeble Minded*, 274 U.S. 200, 47 S. Ct. 584; 71 L. Ed. 1000; 1927.

57 Ibid., 207.

58 See Searle, G. 1979. Eugenics and politics in Britain in the 1930s. *Annals of Science*. 36(2). 159–69, 163–4; Kevles, D. 1986. *In the Name of Eugenics: Genetics and the uses of human heredity*. Berkeley: University of California Press, 47–9.

59 *Association for Molecular Pathology v. Myriad Genetics*, 569 S. Ct. 12–398 (2013), Opinion of Scalia, J.

varieties were thought of more as moving populational targets. Over the decades since mid-century the variety has slowly been ossified as a fixed and discrete unit. Variability has been recast within acceptable boundaries which do not threaten a variety's integrity.[60] In the years since 1980, this conception of variety has in turn been displaced as the product of plant breeding by allegedly fixed identifiable entities around which intellectual property protection could be circumscribed, without the need for maintenance – DNA sequences. Even a basic understanding of biology undermines this fiction: all biological material varies over time, asexually reproducing organisms evolve and PCR amplification – the process of creating enough copies of a DNA sequence to be able to work with – introduces fidelity errors in much the same ways as naturally occurring transcription and translation – the processes which copy DNA in the wild. The law's recent focus on DNA sequences, as though they were static and unchanging over time, degrades the importance of stewardship roles (especially those conducted by small scale farmers in maintaining land races).

A single embodiment of a patented invention is enough to constitute infringement. Likewise, a single plant in isolation or in a mixed population is enough to infringe upon a UPOV registration. When he used Roundup, to check his plants were resistant and to ensure the next generation was homogenously resistant, Bowman was clearly utilising Monsanto's technology. As Stephen Hubicki's chapter in this volume shows, in Europe and Canada, discussions over infringement of patented DNA sequences in those jurisdictions have considered the utilisation of the sequence in the allegedly infringing plant. Even Monsanto's lawyer, Waxman, seemed open to such arguments. On his view, without the application of Roundup,

> There would be inadvertent infringement ... but there would be no enforcement of that ... The farmer wouldn't know, Monsanto wouldn't know, and in any event ... if the farmer doesn't want Roundup Ready technology and isn't using Roundup Ready technology to save costs and increase productivity, the – the royalty value would be zero.[61]

However, the court did not argue on this basis and instead pointed vaguely to planting and harvesting.

Concluding Thoughts

Plant breeding has changed a great deal in the last 150 years. The varying mix of groups who conduct plant breeding – amateurs, farmers, seed firms and publicly funded researchers – has shifted greatly in favour of seed firms over this period.

---

60   Berry, above n 2, 35.
61   *Bowman*, above n 1, Oral arguments, 43, ln. 15–24.

Bowman v. Monsanto Co. exemplifies that shift. Plant breeding is increasingly concentrated in a commercial system that aims to generate profits through increasing quality and yield. At the same time the law pertaining to intellectual property in biological objects has also shifted.[62] The *Bowman v. Monsanto Co.* decision shows that the court is relying on a common legal view of plants and their DNA, which would be alien to most plant breeders, scientific or commercial, throughout the century, as well as alien to the facts of biological reproduction.

One feature of these changing relations – documented here – has been a forgetting of history. What is more, a forgetting that has occurred despite botanists' and plant breeders' historical inclinations for much of the nineteenth and arguably well into the twentieth centuries. The history of plant breeding presented by the American Soybean Association was one which forgot the aims of the free seed schemes and much of government intervention in agriculture over the century – to increase acreage, something such schemes were arguably very successful in doing. A de-historicised view of plant breeding is one that omits the importance of government. This is part of a pattern of undervaluing the importance of public projects.[63] As current economic fashion swings towards *laissez faire* and small government, all the while as we are faced with collective, societal problems, such as climate change, the aims and success of historical public schemes that promoted the good of the nation (albeit clumsily, and unfairly) are worth keeping in view. The importance of such schemes is underscored by them having been undertaken at the height of nineteenth and twentieth century free-trade economics. There are some types of plant breeding that the government has done very well and perhaps government intervention presents a better way to address problems faced by farmers who do not produce enough profit to interest multi-nationals. As economist Thomas Piketty has noted, 'There are nevertheless ways democracy can regain control over capitalism and ensure that the general interest takes precedence over private interests, while preserving economic openness and avoiding protectionist and nationalist reactions.'[64]

The view of making presented by the Supreme Court is one that neglected the historical dimension of biological nature. In focusing on bare DNA, and treating it as a static object, the Supreme Court is colluding in a fiction about the nature of hereditary material that benefits large commercial firms while undermining the importance of stewardship work. This construction of intellectual property works in favour of the large commercial entities that have been so keen to employ its extensive protection.

Comparing history to the current situation clearly demonstrates that (despite what Justice Roberts might think) 'It ain't necessarily so': there is nothing

---

62   On the long-running history of patentable subject matter see, Beauchamp, C. 2013. Patenting nature: A problem of history. *Stanford Technology Law Review.* 16(2), 257–312.

63   Mazzucato, M. 2013. *The entrepreneurial State: Debunking public vs. private sector myths.* London: Anthem.

64   Piketty, T. 2014. *Capital in the Twenty-first Century.* London: Belknap-Harvard, 1.

inevitable about the situation in which we find ourselves today.[65] If plant breeding has taken a particular path of development, to the benefit of particular social groups, the opportunity to change paths, or even turn back is still available. While *Bowman v. Monsanto Co.* illustrates the status quo, of concentration and extensive intellectual property, there are signs in the wake of *Association for Molecular Pathology v. Myriad Genetics* – discussed by Dianne Nicol in this volume – that this might also be a high watermark.

---

65   The title of Richard Lewontin's famously anti-deterministic manifesto: Lewontin, R. 2001. *It ain't Necessarily So: The dream of the human genome and other illusions*. New York: New York Review Books. See also Lewontin's recent piece in *New York Review of Books* asking to whose benefit the new synthetic biology will be put: Lewontin, R. 2014. The new synthetic biology: Who gains? *New York Review of Books*. May. [Online]. Available at: http://www.nybooks.com/articles/archives/2014/may/08/new-synthetic-biology-who-gains/ [accessed 6 May 2014].

Chapter 3

'The Story of a Love Spurned': Monsanto in the United Republic of Soy

Stephen Hubicki

Introduction

In 2005 the Argentinian government decided to suspend negotiations with the world's largest seed producer, Monsanto Company, after the parties failed to reach agreement about the design of a payment system that would allow the company to collect royalties from Argentinian farmers who used Roundup Ready soybeans. As is well known, these soybeans have been genetically modified to incorporate a DNA sequence which, when expressed in plant cells, confers resistance to glyphosate herbicides, the largest selling crop protection product in the world. For Monsanto, the failed negotiations brought to an end a decade-long attempt to recover royalties. Introduced to Argentina in 1996, Roundup Ready soy was the first transgenic crop approved for commercial release in Argentina. Shortly thereafter, Monsanto licensed a number of local manufacturers to incorporate the technology in their seeds.

The enthusiasm with which Monsanto greeted this announcement was tempered by the misfortune that was to befall its Argentinian patent application, filed with the Instituto Nacional de la Propiedad Industrial Argentina (INPI) in 1995. Officially, the INPI rejected Monsanto's application for a 'revalidation' patent – a patent granted in another country which is later 'revalidated' in Argentina for the term of the foreign patent – on the basis of a formality (Monsanto's application was filed more than 12 months after the grant of the relevant foreign patent, which could not, therefore, be revalidated). Monsanto's hopes of having the decision overturned were dashed by the Argentinian Supreme Court in 2000, when it decided that all revalidation patent applications filed after 1 January 1995 were invalid because the World Trade Organisation's (WTO) Agreement on Trade-Related Aspects of Intellectual Property Rights (TRIPS Agreement) does not oblige its members to issue such patents.[1] As Monsanto's revalidation application was filed later that year, the National Patent Administration rejected the application. As a signatory of the 1978 text of the International Union for the Protection of New Varieties of Plants (UPOV Convention), Argentina's system of plant variety rights protection

---

1 See Martin, N. and O'Farrell, M. 2003. Revalidation patents and the Unilever case. *International Review of Intellectual Property and Competition Law*, 34(2), 119–32.

also provided Monsanto with little scope to prevent Argentinian farmers from saving soy seeds.[2] Finally, in January 2004 Monsanto announced that it would cease marketing Roundup Ready soybeans in Argentina because it was not profitable and also terminated its research and development activities in the country.

Monsanto's departure from the Argentinian market brought to an end only the first stage of its commercial battle with the Argentinian government and the country's farmers. The second stage of this contest was waged against Argentina's commodity exporters in the courtrooms of a number of European countries and, ultimately, the Grand Chamber of the European Court of Justice. Monsanto's actions in national and European courts were designed to compel the Argentinian government to accept the introduction of compensation and technology fees at the point of delivery. A bill promised by the Argentinian government at the end of 2004 would have established a 'technology compensation fund' administered by the Department of Agriculture, to which farmers would contribute 0.35 to 0.95 per cent of their sales of soybeans to elevators and exporters. However, the scheme encountered strong resistance from farmer groups, in particular the Federación Agraria Argentina, a group representing a broad section of over one hundred thousand small farmers, and was never passed.[3]

Monsanto's response to this development in Argentina rehearsed a strategy contemporaneously pursued by Monsanto in Brazil. In 1997, farmers in southern Brazil began to sow contraband Roundup Ready soybeans obtained, via Paraguay, from Argentina and dubbed 'soja Maradona' seeds after the Argentinian football hero, Diego Maradona.[4] When it failed to reach an agreement with the Brazilian government about the payment of royalties to use its Roundup Ready soybeans,[5] Monsanto began notifying the country's major grain exporters, including companies which would later become parties in the Argentinian dispute, that they would be liable for patent infringement if they imported soy meal produced in southern Brazil into Europe, and declared that it would begin actively enforcing its patents and collecting royalties, either in Brazil or, if necessary, when the cargo landed

---

2   However, the 1978 text of the *International Union for the Protection of New Varieties of Plants* 815 UNTS 89 (UPOV Convention) Art 5(1) does not authorise the selling of seeds harvested from previous generations. This did not prevent a flourishing black market in harvested soybeans from developing.

3   Newell, P. 2009. Bio-hegemony: The political economy of agricultural biotechnology in Argentina. *Journal of Latin American Studies*, 41(1), 27–57, 47.

4   Continuing the football theme, a new GM soybean recently introduced by Monsanto to Brazil, which is planned for a 2014 release in Argentina, has been dubbed by Argentina's Agriculture Secretary, Lorenzo Basso, 'Ronaldinho'.

5   Monsanto's ability to collect point of sale royalties in Brazil was complicated by the fact that, in 1999, Greenpeace and the Brazilian Consumer Defence Institute obtained an injunction against the government, suspending its authorisation for the commercial release of Roundup Ready soybeans, granted in September 1998, until an environmental impact study was carried out: Osava, M. 2004. Agriculture: Brazil-Monsanto fight heats up. *Global Information Network*, 9 March, 2004, 1.

in Europe.[6] Monsanto began tracking all ships leaving southern Brazilian ports, eventually notifying the exporting company on a particular soybean shipment that local customs action would be taken at the European port. These actions brought about their intended effect:[7] about the same time that Monsanto announced its withdrawal from the Argentinian market, it reached an in principle agreement with southern Brazilian farmers' associations whereby growers agreed to pay a so-called 'end point' or 'point of delivery' royalty for those soybeans grown from seed for which royalties had not been paid.[8] Royalties, or 'indemnification fees', as they were known to Monsanto's Brazilian legal team, are collected on Monsanto's behalf by local elevator operators, who effectively became agents of the company.

Emboldened by the success of this strategy in Brazil, Monsanto began approaching Argentinian soy commodity exporters with a proposal to charge a fee of between US$15 and US$18.75 for each tonne of soy products exported from Argentina.[9] To Argentina, still in the process of recovering from the catastrophic collapse of its economy in 2001, following the government's default on its US$140 billion debt, a recovery that is to a considerable extent underwritten by exports of soybeans, soy meal and soy oil,[10] Monsanto's strategy of attempting to exert pressure on the government by strangling its export market was never likely to be regarded as a satisfactory way of resolving the conflict. Agricultural exports were a key pillar of President Carlos Menem's neo-liberal economic reform agenda in the 1990s, filling the void left by the decline of the manufacturing sector that followed the liberalisation of the Argentinian economy from the late 1970s.[11] Agricultural biotechnology in general, and GM soy in particular, were central to Menem's vision of an export-driven agrarian model of growth. As Peter Newell has noted, the Argentinian economy was already underpinned by a form of agro-hegemony that made it particularly susceptible to a bio-hegemony predicated on

---

6 Bell, D. and Shelman, M. 2006. Monsanto: Realizing biotech value in Brazil. *Harvard Business School Case 507–018*, 8. [Online]. Available at: http://hbr.org/product/monsanto-realizing-biotech-value-in-brazil/an/507018-PDF-ENG [accessed 7 August 2013].

7 However, the move is said to have fomented resentment among Cargill Inc, Archer Daniel Midlands Co and other multinational traders. See Karp, J. and Jordan, M. 2003. Monsanto appears to pressure Brazil over soy piracy. *Wall Street Journal*, 13 June 2003, A. 3.

8 The royalty payable depended on whether the farmer self-declared the use of Roundup Ready soybeans, in which case a royalty of 10 Reals (approximately US$3.40) per metric tonne was payable. If a farmer chooses not to self-declare, the local elevator operator tests for the presence of the glyphosate-herbicide resistance gene; if the test is positive, a royalty of 25 Reals is payable. Over 95 per cent of farmers self-declared in the first year. See Bell and Shelman, above n 6.

9 At the time, Argentinian soy was trading at around US$178 per tonne. See Balch, O. 2006. Seeds of dispute. *The Guardian*, 22 February 2006. [Online]. Available at: http://www.theguardian.com/science/2006/feb/22/gm.argentina [accessed 30 October 2013].

10 Ninety-eight per cent of soy production in 2003 was exported.

11 Newell, above n 3, 31.

biotechnology.[12] Biotechnology represents a 'long term political strategy' that is pivotal to Argentina's national development.

Argentina has embraced this vision with almost unparalleled vigour. After the US, Argentina is the world's largest producer and exporter of GM crops. Argentinian fields produce almost a quarter of the global GM crop, with over eighteen million hectares under cultivation. Billed as 'the success story of the country', soy is Argentina's most extensive GM crop, with the country producing more transgenic soy than any other country, aside from the US and Brazil. Only three years after Roundup Ready soy was introduced in Argentina, three-quarters of Argentina's soy production was genetically modified.[13] From the time of its introduction in 1996 until the 2004/2005 planting season, the number of hectares allocated to soy production rose from 10,000 to more than fourteen million.[14] In 2003, GM soy accounted for half of Argentina's total agricultural production and a quarter of the country's total income from exports. By 2006, more than half of the arable land in Argentina was sown with transgenic soy.[15] In the 2006/2007 planting season, sixteen million hectares were sown with transgenic soy and production reached its historical record of forty-eight million tonnes.[16] By 2010, Roundup Ready soy accounted for almost 90 per cent of the national planted area.[17]

The exponential growth in transgenic soy production in Argentina has had numerous contributors: the 'mad cow' epidemic which struck the European market in the 1990s resulted in the increasing use of soy meal in animal feeds, in preference to meat and bone meal, as authorities attempted to halt transmission of the disease; the devaluation of the peso (by 70 per cent) in 2002; the emergence of China as a major export market[18] and, of course, the absence of patent protection for Roundup Ready soy in Argentina. This latter contributor has allowed Argentinian producers to save seeds from one crop cycle and use them to produce another crop (an activity denied to farmers where the technology is subject to patent protection), whilst also stimulating the growth of a black market in Roundup Ready soy seeds. According to unofficial estimates, between 70 and upwards of 80 per cent of the seeds sown by Argentinian soy farmers are 'bolsa blanca' seeds,[19] seeds obtained on the black market or saved from previous generations, which take their name

---

12   Ibid., 35–6.

13   Ibid., 31.

14   Correa, C. 2006. La disputa sobre soja transgénica. Monsanto vs Argentina. *Le Monde Diplomatique/El Dipló*, April 2006.

15   Viollat, P.-L. 2006. Argentine, un cas d'école. *Le Monde Diplomatique*, 53(625), April 2006, 22.

16   Binimelis, R., Pengue, W. and Monterroso, I. 'Transgenic treadmill': Responses to the emergence and spread of glyphosate-resistant Johnsongrass in Argentina. *Geoforum*, 40(4), 623–33, 625.

17   Anon, 2010. Argentina: Seed exports top US$200 mil. *El Cronista*, 21 June 2010.

18   More than half of Argentina's soy production is exported to Europe, with the remainder going to China.

19   Newell, above n 3, 45.

from the unlabelled white pouches in which they are distributed. At the time Monsanto withdrew from the Argentinian market, no more than 20 per cent of soybeans cultivated in Argentina were obtained from authorised dealers, down from 50 per cent in 1996.[20]

By the time negotiations with Monsanto broke down, the Argentinian economy was characterised by a 'near total dependence' on transgenic soy.[21] The agreement for repayment of the country's debt, struck by the government with the International Monetary Fund in 2003, is based on an assumption that 40 per cent of the total repayment would come from taxes on exports of soy and its derivatives.[22] Moreover, more than half of the country's tax revenue is derived from tax applied to exports of soy products,[23] further undermining the ability of small farmers, in particular, to commit to further royalties.[24]

The Argentinian government criticised Monsanto's negotiating position for being insensitive to the country's situation. Emerging from the failed negotiations with Monsanto, Argentina's then Agriculture Secretary, Miguel Campos, told the assembled media, 'Monsanto looks at Argentina as if it were just an Excel spreadsheet.'[25] Earlier, Campos had reminded Monsanto of his country's 'audacity' in approving the commercial planting of Roundup Ready soy. 'The great beneficiary of this', he offered, 'has been Monsanto. Argentina has been the launching point for the use of this technology on the continent. This has allowed Monsanto to make advances in other countries.'[26] Indeed, a report published in The Guardian in 2004 characterised the imbroglio as 'the story of a love spurned'.[27] Argentina's location and climate made it an ideal winter nursery for testing and multiplying promising lines, providing Monsanto with two growing seasons a year and shortening the product development pipeline.[28] Moreover, Argentina's support for GM crops is unsurpassed in the region,[29] a point which Campos was keen to emphasise when he accused Monsanto of having forgotten 'everything Argentina has done to advance biotechnology', a pointed reminder of the country's support for the case initiated by the US (along with Argentina and Canada) in the World Trade

---

20   Turner, T. 2005. Argentina to fight Monsanto in Court, suspend soybean talks. *Dow Jones Online*, 8 July 2005.

21   Newell, above n 3, 37.

22   Joensen, L., Semino, S. and Paul, H. 2005. *Argentina: A case study on the impact of genetically engineered soya*. Report prepared for the Gaia Foundation, 10. [Online]. Available at: http://www.econexus.info/publication/argentina-case-study-impact-genetically-engineered-soya [accessed 29 October 2013].

23   Binimelis *et al.*, above n 16, 625.

24   According to a spokesman for the Argentine Agrarian Federation, which represents up to 60,000 farmers, 'we could never accept this. It would be terrible': Turner, above n 20.

25   Ibid.

26   Ibid.

27   Balch, above n 9.

28   Bell and Shelman, above n 6, 11–12.

29   Newell, above n 3, 37.

Organisation in 2003, challenging the European Union's *de facto* moratorium on the commercial approval of GM crops, despite the potential for such action to sour relations with Europe, the primary destination for its soy meal. 'If there has been a real strategic ally of biotechnology', Campos continued, 'it has been Argentina.'[30]

In 2006, after failing to reach agreement with the Argentinian government and European importers, who rejected Monsanto's proposal to charge a fee of between US$15 and US$18.75 for each tonne of soy products exported from Argentina,[31] Monsanto made good on its threats, initiating proceedings for infringement of its European patent against importers in several European countries, including Denmark, Spain, the Netherlands, and the United Kingdom.[32] The patent in suit broadly relates to a method of producing GM plants that are tolerant of Monsanto's proprietary 'Roundup Ready' herbicides and includes claims to: an isolated DNA sequence (a Class II 5-enol-pyruvylshikimate-3phosphate synthase (EPSPS) enzyme) which, when transferred to and expressed in plant cells, confers resistance to glyphosate herbicides; and, glyphosate-resistant plant cells containing the DNA sequence.

Trace quantities of the claimed DNA sequence were found in the soy meal imported into Europe, which was processed in Argentina from plants that embodied Monsanto's invention and were grown there. Monsanto alleged that the presence of these DNA sequences in the imported soy meal infringed the claims

---

30   Argentinian government officials also accused Monsanto of having actively encouraged the uptake of Roundup Ready soy by Argentinian farmers. According to one official within the Agriculture Ministry, the use of seeds for cultivation is done with the direct support of biotechnology companies for whom it is 'a logical firm strategy', presumably because it leads to further sales of glyphosate-herbicides. Indeed, initially most of Monsanto's profits in Argentina were generated from sales of its Roundup herbicide. The expiration of its patent on the herbicide in 2000, after which Monsanto encountered stiff price competition from cheaper Chinese imports, as well as growing discontent among North American soy farmers, aggrieved at paying more than double the price Argentinian farmers paid for soy seed (with most of the difference in price being attributable to the technology use fee which North American farmers had to pay to use Monsanto's seeds), compelled the company to seek to recover a greater percentage of the profits earned by Argentinian farmers for the use of its seeds. See Newell, ibid., 42; Laursen, L. 2010. How green biotech turned white and blue. *Nature Biotechnology*, 28(5), 393–5, 394; Lambrecht, B. 2000. Difference in Monsanto's seed prices angers US farmers: Bag of soybeans costs $9 in Argentina, $21.50 in US. *St. Louis Post*, 5 March 2000, E 1. According to this report, North American farmers pay a $6.50 technology use fee as part of every seed purchase.

31   At the time, Argentinian soy was trading at around US$178 per tonne: Balch, above n 9. Under the proposal, exporters would have paid Monsanto US$3 per tonne to export the soy meal and US$15 per tonne upon importation into Europe: Heath, C. 2009. The scope of DNA patents in the light of the recent Monsanto decisions. *International Review of Intellectual Property and Competition Law*, 40(8), 940–56, 942.

32   Barry, G., Kishore, G. and Padgette, S. 1993. *Glyphosate Tolerant 5-Enolpyruvylshikimate-3-Phosphate Synthases*. European Patent No 0546090, 16 June 1993.

relating to isolated DNA sequences, as well as the method claims for producing GM soy plants incorporating these DNA sequences (on the basis that the soy meal was the direct product of the patented process). Recalling Richard Dawkins' idea of DNA as indifferent, purposeless matter ('DNA neither cares nor knows; [it] just is'),[33] Monsanto argued that the purpose for which the DNA sequence was prepared and operationalised was irrelevant to the question of infringement; all that matters is that the claimed DNA sequences continue to persist in some form, utile or otherwise.

The primary question before each national court was whether the 'presence alone [of the claimed DNA sequence] is sufficient to constitute infringement of Monsanto's European patent when the soy meal is imported into the European Community'.[34] Monsanto alleged that importation of the soy meal constituted infringement under both national patent legislation – in each country the rights to import and to use the patented invention are exclusive to the patentee – and pursuant to Article 9 of the European Biotechnology Directive. Article 9 provides that:

> the protection conferred by a patent on a product containing or consisting of genetic information shall extend to all material, save as provided in Art 5(1), in which the product is incorporated and in which the genetic information is contained and performs its function.[35]

Monsanto argued that this provision extends protection to products that incorporate the genetic material and in which the genetic material is able to fulfil a function. It is not necessary, in Monsanto's view, to show that the genetic information is actively performing its function at the time of the infringement.[36] Instead, Monsanto argued that liability for infringement under Art 9 may be established by showing that the genetic information has performed its function in the past, or may perform its function in the future: it is enough that the function of the genetic information has been, or will be, effectuated.[37]

---

33   See Dawkins, R. 1995. *River Out of Eden*, London: Weidenfeld & Nicholson, 133. See also Sommers, T. and Rosenberg, A. 2003. Darwin's nihilistic idea: Evolution and the meaninglessness of life. *Biology and Philosophy*, 18(5), 653–68.

34   *Case C-428/08, Monsanto Technology LLC v. Cefetra BV* [2010] E.C.R. I-6768 (Advocate General Mengozzi's Opinion); I-6800 (Decision of the Grand Chamber), [22].

35   Article 5(1) provides that 'the human body, at the various stages of its formation and development, and the simple discovery of one of its elements, including the sequence or partial sequence of a gene, cannot constitute patentable inventions.' In the context of Art 9, the effect of Art 5(1) is to confirm that claims to human gene sequences do not embrace those genes in their natural state in the human body.

36   F.B. 2009. *Monsanto Technology LLC v. Sesostris SAE* – 'Roundup Ready Spain'. *International Review of Intellectual Property and Competition Law*, 40(2), 233–237, 235.

37   C.H.   2009.   *Monsanto   Technology   v.   Cefetra*,   Argentina   *et   al.*   – 'Roundup Ready'. *International Review of Intellectual Property and Competition Law*, 40(2) 228–32, 233 ([4.17.3]).

Except for The Hague District Court, each national court rejected Monsanto's claim.[38] As a consequence, Monsanto was unable to establish infringement in any of the four countries in which proceedings were commenced, despite the fact that analyses of the composition of the meal revealed that traces of the claimed DNA sequence were present. It was also established that the seeds from which the soy meal was derived had been planted in Argentina and that these plants contained the patented DNA sequence. As Justice Pumfrey noted in the United Kingdom litigation, the plants from which the meal was derived could very well be viewed as the 'lineal descendants' of the original transformed plant[39] and this genealogy could be traced through to the fragments of DNA found in the imported meal. Monsanto was therefore able to establish that the patented DNA sequence was present in both the imported meal and the plants from which the meal was derived and that the meal could legitimately be described as 'the ultimate product of the original transformation of the parent plant'.[40] However, neither of these facts was sufficient to establish infringement in any of the national courts.

In contrast to the assured manner in which Spanish courts dispensed with Monsanto's arguments, The Hague District Court was perplexed by the potential conflict between the operation of domestic patent law and the Directive. Monsanto argued that Arts 8 and 9 did not apply to the patent in suit,[41] or, in the alternative, that the principle of 'absolute protection' continues to exist alongside the specific protection provided by Art 9. That is, Art 9 sets only a minimum standard of protection. The court opined that 'there seems to be reason to assume that the Directive does not alter the absolute protection afforded by s 53a(3) Patent Act, and rather provides for a minimum protection,'[42] Support for this interpretation

---

38   Due to a transitional period, the Directive had not yet commenced operation in the United Kingdom at the time the action was heard. The English Patents Court accordingly considered the question of infringement from the perspective of national patent law only.

39   *Monsanto Technology LLC v. Cargill International SA* [2008] F.S.R. 153, 173. Likewise, the Commercial Court of Madrid found that the imported meal had been produced from soybeans containing the patented DNA sequence: Roundup Ready Spain, above n 36, 236.

40   *Monsanto Technology LLC v. Cargill International SA* [2008] F.S.R. 153, 173.

41   This argument was based on the dubious reasoning that the imported soy meal is not (biological) material within the meaning of Arts 8 and 9, which, accordingly, are inapplicable: Roundup Ready, above n 37, 230 ([4.15.1]). Not surprisingly, this argument was rejected by the court, which emphasised that these provisions should be construed as relating to the biological material for which the patent was granted, not the allegedly infringing material: 231 ([4.20]).

42   Ibid., 231–2 ([4.27]). On the other hand, Christopher Heath, a member of the Boards of Appeal of the EPO, has argued that, given that the Directive is a compromise between 'widely different perceptions on how biotechnological inventions should be protected', it should not be viewed as setting a minimum standard of protection. Thus, the point of reference for determining the scope of biotechnological patents is not the absolute product protection contained in national patent law, but Arts 8–11 of the Directive.

was to be found in the wording of Art 9, which uses the verb 'extends to' and not, for example, 'is limited to', and Arts 3(2) and 5(2), which confirm the patentability of isolated DNA.[43]

In view of this conundrum,[44] the Hague District Court referred the following four questions to the Court of Justice of the European Union for a preliminary ruling:[45]

1. Can Art 9 be invoked in circumstances where a product (the DNA sequence) forms part of a material imported into the European Union and is not performing its function at the time of the alleged infringement, but did perform that function previously or could possibly again be able to perform that function after it has been isolated from that material and introduced to the cell of an organism?

2. Does Art 9 prevent domestic patent law from additionally conferring absolute protection on the product (the DNA sequence) as such, regardless of whether the DNA performs its function, or is the protection provided by Art 9 exhaustive in the situations where the product consists of, or contains, genetic information, and the product is incorporated in material which contains the genetic information?

This means that if a patent claim that is relied on in an infringement action falls within one of the categories of Arts 8 or 9, the scope of protection is exhaustively determined by those provisions: Heath, above n 31, 945. This interpretation is supported by the CJEU, which noted that, in the interest of avoiding barriers to trade within the internal market, the Community legislature 'intended to effect a harmonisation which was limited in its substantive scope, but suitable for remedying the existing differences and preventing future differences between Member States in the field of protection of biotechnological inventions': Case C-428/08, above n 34, I-6809, [55]–[56].

43   Roundup Ready, above n 37, 231–2 ([4.27]). Article 3(2) provides that biological material which is isolated from its natural environment or produced by means of a technical process may be the subject of an invention even if it previously occurred in nature, whilst Art 5(2) provides that an element isolated from the human body or otherwise produced by means of a technical process, including the sequence or partial sequence of a gene, may constitute a patentable invention, even if the structure of that element is identical to that of a natural element.

44   The court was also concerned that, if the Directive did not permit absolute protection, this would lead to 'the inconsistent situation that even an isolated DNA, as long as it is not further processed, would not be included in the scope of protection', presumably on the basis that the DNA is not capable of performing its function whilst it is isolated. However, this fear is unfounded: if the alleged infringement concerns isolated DNA then Art 9 does not apply since, by definition, it has not been incorporated into any other material.

45   The preliminary ruling procedure entitles a national court to refer questions of interpretation regarding European law to the CJEU. The 'preliminary' nature of the procedure arises from the fact that proceedings in a national court are stayed pending the resolution by the CJEU of the questions referred to it. The CJEU's decision is binding on all national courts within the European Union.

3. Does the fact that a patent was granted prior to the adoption of the Directive, and that absolute protection was provided under domestic law, affect the answer to question 2?
4. Do Arts 27 and 30 of the TRIPS Agreement affect the interpretation given to Art 9?

In this chapter, I examine the answers provided by the CJEU to these questions, and the largely critical reaction to the court's decision. Critics have characterised the decision as an affront both to received wisdom regarding the scope of claims to isolated DNA sequences, and the legislative intent of the drafters of the Directive. Here, I offer a counter-narrative and suggest that the CJEU's decision is explicable in terms of both general principles of patent law, and the negotiating history of the Directive and the World Intellectual Property Organisation (WIPO) Committee of Experts, from which Art 9 is ultimately derived. I commence with an overview of the CJEU's decision, followed by an analysis of the major criticisms of the decision. I then examine the veracity of these criticisms against the work of the WIPO Committee of Experts, which laid much of the groundwork on which the Directive is established, as well as the negotiating history of the Directive.

The Decision of the CJEU

The First Question

The CJEU answered questions 1, 3 and 4 in the negative.[46] In respect of the first question, the CJEU observed that Art 9 had been drafted in the present tense. This implies that the genetic information must be performing its function 'at the present time and in the actual material in which the DNA sequence containing the genetic information is found'.[47] It follows from this that the protection provided by Art 9 is not available where the genetic information has ceased to perform the function 'it performed in the initial material from which the material in question is derived'.[48] It would appear that 'function' is not to be interpreted in a general or biological sense, but with reference to the function of the DNA sequence disclosed in the specification.[49] Hence the function of the genetic information in the present case 'is performed when the genetic information protects the biological material in which

---

46   Case C-428/08, above n 34, I-6800, [22].
47   Ibid., I-6802, [35].
48   Ibid., I-6802, [38].
49   This accords with the position adopted by the Human Genome Organisation Intellectual Property Committee. 2003. *Statement on the Scope of Gene Patents Research Exemption, and Licensing of Patented Gene Sequences for Diagnostics*, [2]. [Online]. Available at: www.hugo-international.org/img/ip_gene_2003.pdf [accessed 3 November 2013].

it is incorporated against the effect, or the foreseeable possibility of the effect, of a product [i.e. a herbicide] which can cause that material to die'.[50] The Court found that the patented genetic information was no longer capable of performing this function. Even if it is customary to spray soy meal with herbicide, the genetic information would still be unable to perform the function of 'protect[ing] the life of the biological material containing it' because the information had been rendered 'dead material' by processing.[51] Nor was it possible to circumvent this interpretation by proving that the genetic material might be capable of once again performing its function. To allow protection to be revived by showing that the genetic information could be extracted from the soy meal and transferred to another organism, in which it could once again perform its function, would diminish the effectiveness of Art 9 since 'in principle' this potentiality could 'always be relied on'.[52]

The Court also resisted Monsanto's attempts to escape the operation of Art 9 by invoking the principle of 'absolute protection'. It is a generally accepted principle that the scope of protection conferred by a European patent in respect of a product is 'absolute' in the sense that the owner of the patent may prevent anyone from making the product by any method (including methods not described by the patentee), and from using the product for any purpose whatsoever (including purposes the patentee was unaware of). Others may still obtain a patent for a new and non-obvious way of making or using the product, but they cannot work their invention without the authorisation of the patentee of the product (and *vice versa*).[53] Whilst the Court acknowledged that the scope of protection for a DNA sequence 'is indeed absolute under the applicable national law', it held that this principle was repugnant to Art 9. If sustained, Monsanto's argument would render Art 9 otiose because the '[p]rotection accorded formally to the DNA sequence as such would necessarily in fact extend to the material of which it formed a part, as long as that situation continued'.[54]

The Court buttressed this interpretation of Art 9 on Art 5(3) and Recitals 23 and 24.[55] As the provisions make 'the patentability of a DNA sequence subject to indication of the function it performs', the Directive 'must be regarded as not according any protection to a patented DNA sequence which is not able to perform

---

50 Case C-428/08, above n 34, I-6804, [36].
51 Ibid., [37].
52 Ibid., I-6805, [40].
53 In English law, this is one consequence of the principle of absolute liability.
54 Case C-428/08, above n 34, I-6806, [47].
55 Article 5(3) provides that the industrial application of a sequence or a partial sequence of a gene must be disclosed in the patent application. Recital 23 provides that a mere DNA sequence without an indication of a function does not contain any technical information and is therefore not an invention, whilst Recital 24 provides that, where the invention relates to a gene or a partial sequence of a gene which is used to produce a protein, the industrial applicability criterion is satisfied by specifying which protein or which part of a protein, or what function it performs.

the specific function for which it is patented'.[56] In other words, there is a direct relationship between patentability and scope of protection.[57] Accordingly, the answer to the first question is that:

> Article 9 of the Directive must be interpreted as not conferring patent right protection in circumstances such as those of the case in the main proceedings, in which the patented product is contained in the soy meal, where it does not perform the function *for which it was patented*, but did perform that function previously in the soy plant, of which the meal is a processed product, or would possibly again be able to perform that function after it had been extracted from the soy meal and inserted into the cell of a living organism.[58]

Thus, in answering the first question, the CJEU essentially formed a view that the *raison d'être* of patents relating to genetic information is the disclosure of the function of that information. Absent this disclosure, the genetic information is not patentable. Likewise, once the genetic information ceases, or loses the capacity, to perform this function, so too does protection: what is incapable of being patented is incapable of being protected.

The Second Question

The CJEU held that the Directive, in particular Art 9, does not merely set a minimum level of patent protection for biotechnological inventions but is exhaustive with respect to the matters that are addressed by the Directive. The Court predicated this interpretation upon what it perceived to be the intentions of the Community legislature when drafting the Directive. According to the Court, the legislature's intention was to 'effect a harmonisation which was limited in its substantive scope, but suitable for remedying the existing differences and preventing future differences between Member States in the field of protection of biotechnological inventions'.[59] This interpretation was supported by a number of Recitals, which emphasised that the Community legislature's overarching concern in introducing the Directive was to remove barriers to trade. Such barriers are

---

56 Case C-428/08, above n 34, I-6805, [44].

57 It is by no means clear that this interpretation is correct. The relevant provisions require only that the applicant indicate or specify what the function of the patented DNA sequence is. There is nothing else in the Directive that might be taken to indicate that the relevant claims must be so limited. In its absence, to interpret the scope of a claim to a DNA sequence *per se* as limited to the function(s) specified in the body of the specification is to read into the claim a limitation that is not there. It may be that, as is the case generally, the obligation to disclose the industrial application of a DNA sequence is only a threshold requirement.

58 Case C-428/08, above n 34, I-6806–7, [50].

59 Case C-428/08, above n 34, I-6808, [55].

created by differences in the law and practices of individual Member States, which can act as a disincentive to trade between Member States (to the detriment of the industrial development of biotechnological inventions) and compromise the proper functioning of the internal market. The Court held that the purpose of the Directive would be negated by a 'minimalist harmonisation approach' that would allow Member States to provide protection above and beyond that provided by the Directive. This would have the effect of entrenching or creating differences with respect to such matters between Member States, 'thereby fostering barriers to trade'.[60] Accordingly, 'insofar as the Directive does not accord protection to a patented DNA sequence which is not able to perform its function, the provision interpreted precludes the national legislature from granting absolute protection to a patented DNA sequence as such, regardless of whether it performs its function in the material containing it.'[61]

The Third Question

The Court also held that the Directive applies to patents that were granted prior to the commencement of the Directive. The Court reaffirmed settled case law to the effect that newly introduced Community legislation applies immediately to 'the future effects of a situation which arose under the old rule'.[62] Thus, where a patent has been granted by a Member State and, according to the national law of that Member State, the invention claimed in the patent enjoys absolute protection, the patentee is unable to rely upon national patent law insofar as that law is inconsistent with the Directive. Referring once more to the paramount purpose of the Directive, the Court stated that any other interpretation might result in the emergence of differences in the scope of protection between Member States and compromise the substantive harmonisation desired by the Community legislature.

The Fourth Question

Finally, the Court held that although it is necessary to 'supply an interpretation in keeping with the TRIPS Agreement', it confirmed that 'the provisions of the TRIPS Agreement are not such as to create rights upon which individuals may

---

60   Ibid., I-6808–9, [59].

61   Ibid., I-6810, [62]. This interpretation was unaffected by Art 1(1), which provides that 'Member States shall protect biotechnological inventions under national patent law' and, 'if necessary, adjust their national patent law to take account of the provisions of this Directive'. According to the Court, Art 1(1) requires Member States, where necessary, to adjust their national patent law to take account of the provisions of the Directive, 'that is, in particular, those effecting exhaustive harmonisation'.

62   Ibid., I-6811, [66] (referring to *Case C-334/07 P, Commission v. Freistaat Sachsen* [2008] E.C.R. I-9465).

rely directly before the courts by virtue of European Union law.'[63] In any event, the Court opined that Arts 27 and 30 of the TRIPS Agreement are concerned with patentability and exceptions to the rights conferred by a patent respectively. Article 9, on the other hand, pertains to the scope of protection conferred by a patent on its holder. Further, the Court rejected the notion that, assuming that the reference in Art 30 of the TRIPS Agreement to 'exceptions to the rights conferred' could be construed as encompassing not only exclusions to the rights of patentees but also limitations upon those rights, Art 9 could not be considered either 'to conflict unreasonably with a normal exploitation of the patent', nor '"unreasonably prejudice the legitimate interests of the patent owner, taking account of the legitimate interests of third parties", within the meaning of Art 30 of the TRIPS Agreement'.[64] The interpretation of Art 9 of the Directive is therefore unaffected by Arts 27 and 30 of the TRIPS Agreement.

Analysis

Anxious to downplay the significance of the CJEU's decision and allay the concerns of investors, in the days following the decision Monsanto published on its website a statement offering an upbeat assessment of the decision, emphasising that it is 'very limited and will have no impact on Monsanto's global Roundup Ready soybean business. The applicability of the Directive to whole soybeans containing Monsanto's Roundup Ready trait, where the gene clearly maintains its functionality, was not at issue'.[65]

However, other commentators have proffered a decidedly less sanguine interpretation of the decision. For some, the decision is symptomatic of a general antipathy towards biotechnology and GMOs in Europe, whilst others have questioned the competency of the CJEU in matters pertaining to patent law.[66] In this regard, a number of commentators have argued that the decision is unfaithful to the legislature's intent, which they suggest is to 'extend' the scope of protection for genetic information beyond that provided by national patent law, as well as to clarify the circumstances in which a patentee's rights in relation to an invention

---

63   Ibid., I-6812, [71–2].

64   Ibid., I-6813, [76].

65   Monsanto. 2013. *Argentine soy meal imports*. [Online]. Available at: http://www.monsanto.com/newsviews/Pages/argentine-soy-meal-imports.aspx   [accessed   4 October 2013].

66   'The present case shows the consequences which arise once the European Union gets their hands on patent matters': Storz, U. and Huettermann, A. 2010. *Monsanto soy bean patent cases – A paradigm shift gathering in case the ECJ takes over patent jurisdiction*, *Les Nouvelles*. [Online]. Available at: http://www.lesi.org/les-nouvelles/les-nouvelles-online/2010/september-2010/2011/05/02/monsanto-soy-bean-patent-cases---a-paradigm-shift-gathering-in-case-the-ecj-takes-over-patent-jurisdiction [accessed 28 March 2014].

consisting of, or containing, genetic information are exhausted. This section examines each of these criticisms in turn.

The argument that the CJEU's interpretation of Art 9 is inconsistent with the legislative intent encounters several difficulties, not the least of which is a statement made by a European Commission official confirming that the CJEU's decision 'backs up the Commission's interpretation of the Biotech Directive'.[67] More generally, given the tumultuous history of the Directive, any attempt to divine the legislative intent embodied in the various provisions of the Directive must be regarded as an ambitious project, to say the least. As the myriad different interpretations suggested by commentators attest,[68] this is particularly true of Art 9, which was revised on no less than three occasions during the negotiation of the Directive (see Table 3.1) for reasons that are far from clear. Further, the Explanatory Memorandum to the Original Proposal is the only document produced during the negotiation of the Directive to elaborate on the meaning of Art 9 (then Art 13); no meaningful explanation was offered for the wording of any of the subsequent revisions.

Nevertheless, the negotiating history of the Directive does provide some guidance in relation to the drafters' intentions. The working documents of the WIPO Committee of Experts on Biotechnological Inventions and Industrial Property are particularly instructive in this regard. As explained below, the provisions of the Original Proposal in general, and Art 9 in particular, are indebted to the work of this Committee.[69]

Prolegomenon to the Directive: The WIPO Committee of Experts on Biotechnological Inventions and Industrial Property

In September 1983 the International Association for the Protection of Intellectual Property (Association Internationale pour la Protection de la Propriété Intellectuelle or AIPPI), a key figure in the development of international industrial property law and policy from the late nineteenth century onwards, 'instructed' WIPO to investigate the current state of protection for biotechnological inventions, by patent or any other means, and the possible means of providing for industrial property

---

67 Miller, J. 2010. Monsanto loses case in Europe over seeds. *Wall Street Journal,* 7 July 2010, B. 1.

68 See, for example, Kamstra, G. 2002. *Patents on Biotechnological Inventions: The EC Directive.* London: Sweet & Maxwell, 2002, 50–51; Llewellyn, M. and Adcock, M. 2006. *European Plant Intellectual Property,* Portland: Hart, 2006, 378–80; Krauss, J. and Takenaka, T. 2011. A special rule for compound protection for DNA-sequences – Impact of the ECJ 'Monsanto' decision on patent practices. *Journal of the Patent and Trademark Office Society,* 93(2), 189–206.

69 See Straus, J. 2005. AIPPI and the protection of inventions in plants – Past developments, future perspectives. *International Journal of Industrial Property and Copyright Law,* 20(5), 600–621, 601.

protection for such inventions. To assist in this task, the International Bureau of WIPO convened a Committee of Experts on Biotechnological Inventions and Industrial Property, comprised of representatives from 27 nations (including the highest ranking patent official in each country),[70] five international government agencies,[71] and 13 private international organisations.[72]

The Committee met on four occasions between 1984 and 1988. In addition to the objectives set for them by AIPPI, the Committee also considered the possibilities for the improvement of patent protection for biotechnological inventions (including the establishment of treaty or model provisions).[73] Prior to its third meeting in 1987, the International Bureau circulated two questionnaires – one addressed to governments and intergovernmental organisations, the other to non-governmental organisations – concerning the availability, and scope, of industrial property protection for biotechnological inventions under national law.[74] Of particular

---

70   Austria, Belgium, Brazil, China, Denmark, the Dominican Republic, Egypt, Finland, France, (the Federal Republic of) Germany, Hungary, Indonesia, Ireland, Italy, Japan, Madagascar, the Netherlands, Portugal, New Zealand, Norway, Saudi Arabia, the Soviet Union, Spain, Sweden, Switzerland, the United Kingdom and the United States of America. The number of countries represented on the Committee of Experts fluctuated with each meeting.

71   The United Nations Conference on Trade and Development, the World Health Organisation, the Commission of European Communities, the European Patent Organisation and the International Union for the Protection of New Varieties of Plants.

72   The Association of Plant Breeders of the European Economic Community, the Committee of National Institutes of Patent Agents, the European Federation of Agents of Industry in Industrial Property, the Institute of Professional Representatives Before the European Patent Office, the International Association for the Advancement of Teaching and Research in Intellectual Property, the International Association of Plant Breeders for the Protection of Plant Varieties, the International Association for the Protection of Industrial Property, the International Chamber of Commerce, the International Federation of Industrial Property Attorneys, the International Federation of the Seed Trade, the Union of European Practitioners in Industrial Property, the Union of Industries of the European Community and the World Federation for Culture Collections.

73   World Intellectual Property Organisation. 1984. *Industrial property protection of biotechnological inventions: Memorandum prepared by the International Bureau*, BIOT/CE/I/2. Geneva: WIPO. This memorandum was prepared as a basis for discussion at the Committee's first meeting held in November 1984, where it received 'general approval'. At this meeting the Committee directed the International Bureau to prepare a study along the lines described in the memorandum, which would serve as a basis for further discussion at its second meeting in February 1986. In November 1985, the International Bureau produced a report in which a more detailed treatment was given of the issues identified in its earlier memorandum. See World Intellectual Property Organisation. 1985. *Industrial property protection of biotechnological inventions: Report prepared by the International Bureau*, BIOT/CE/II/2. Geneva: WIPO.

74   World Intellectual Property Organisation. 1987. *Annexes to document BIOT/CE/III/2*. Geneva: WIPO. The questionnaire addressed to governments and intergovernmental organisations was exegetical, whilst the questionnaire prepared for non-governmental organisations was normative.

relevance to Art 9 are two questions found in the questionnaire addressed to governments and intergovernmental organisations. The first of these questions (2.1.3) asked, in relation to a hypothetical claim for a new DNA sequence of a particular structure coding for the production of a particular polypeptide, whether the production and/or sale of a microorganism containing the DNA and obtained by incorporating the DNA into it, was an infringement of the claim to the DNA under domestic law. The second question (2.1.5) asked, in respect of a hypothetical claim for a new genetically engineered plant cell, whether the regeneration of a plant from the cell and/or the sale of that plant infringes the claim to the cell under national law. According to the accompanying Explanatory Memorandum, these questions address two issues: first, 'the extension of protection of products produced by *using* a patented process or product';[75] and, second, 'whether the presence of a protected product in a more complex material (where its properties are used and are of importance for the value of the complex material) means that the complex material is covered by the claim'.[76]

In respect of the first question, a majority of countries indicated that a claim to the DNA sequence would be infringed in those circumstances.[77] Notably, the United Kingdom and the US were among those countries that were uncertain of the position under their respective national laws.[78] In the United Kingdom, only a little over a decade earlier the Banks Committee had recommended that '[a] process claim or an apparatus claim shall not be infringed by the sale or use of a product in the manufacture of which the process or apparatus was used.'[79] This was the considered opinion of eminent British patent lawyers at the time.

---

75   World Intellectual Property Organisation. 1986. *Questionnaire concerning the legal protection of biotechnological inventions*, BIOT/Q1. Geneva: WIPO, Chapter II, 40–43 (emphasis supplied).

76   WIPO, Annexes to Document BIOT/CE/III/2, above n 74, 42 (emphasis supplied). The explanatory memorandum is addressed only to question 2.1.3 but the Report in which the responses to these questions is discussed confirms that this issue is common to both questions: World Intellectual Property Organisation. 1987. *Industrial property protection of biotechnological inventions: Revised report prepared by the International Bureau*, BIOT/ CE/III/2. Geneva: WIPO, 31.

77   The countries were Austria, France, the Federal Republic of Germany, Ireland, Italy, Japan, the Netherlands, Spain, Sweden and Switzerland. The Federal Republic of Germany expressed doubts about whether the sale of a product containing the patented genetic information would amount to infringement under national law.

78   WIPO, Revised Report BIOT/CE/III/2, above n 76, 31. The other countries to express doubts about whether infringement takes place in these circumstances under national law were Denmark, New Zealand and Norway. Finland and Portugal indicated that no infringement occurs under national law in either of these circumstances.

79   See Board of Trade. 1970. *The British patent system: Report of the committee to examine the patent system and patent law*, Cmnd. 4407. London: HMSO, 85.

The responses to the second scenario were more equivocal. Only three countries indicated that the claim would be infringed both by manufacture and sale of a plant containing the patented cell, whilst four other countries thought that infringement occurred in relation to manufacture, but not sale, of the plant.[80] A majority of countries, again including both the United Kingdom and the US, remained uncertain as to whether either act amounted to infringement under national law.[81]

Based upon the responses it received to these and the other 109 questions contained in the questionnaire, the International Bureau developed a series of 19 'Suggested Solutions' to the issues raised by each question, with the intention of providing 'guidance for future legislation and practice in industrial property offices and court proceedings'.[82] Notwithstanding their differences, the responses to questions 2.1.3 and 2.1.5 were consolidated into Suggested Solution 13, which is virtually identical to the first iteration of Art 9 (Art 13 of the Original Proposal):

> It appears desirable that patent protection for a product that consists of, or contains, particular genetic information as an essential characteristic of the invention shall extend to any product (the second product) containing the patented product or obtained from such product which contains the said genetic information and in respect of which the said genetic information is of essential importance for the industrial applicability of the said second product.[83]

It is apparent from the Explanatory Memorandum accompanying the questionnaire that this proposal was intended to capture the idea that 'the presence of a protected product' – a DNA sequence and a transformed plant cell, respectively – 'in a more complex material' – a microorganism and a plant respectively containing the protected polypeptide and cell – 'means that the complex material is covered by the claim', provided that the *properties* of the protected product are *used* and are of *importance for the value* of the complex material.[84] It is notable that this idea is

---

80   Finland, Italy and Spain indicated that infringement occurs in this situation under national law, whilst Denmark, France, the Netherlands and Switzerland considered that a clear case of infringement is made out only in relation to regeneration of the plant from the patented cells, but not in respect of the sale of plants containing the patented cells.

81   Hungary, Ireland, New Zealand, Norway, Portugal and Sweden are the other countries that were uncertain about whether these acts constitute infringement under national law. Austria discerned no infringement in either instance, whilst the Federal Republic of Germany indicated that infringement would not arise where the patented plant cell was purchased from the patent owner. Japan did not provide a response to this question.

82   World Intellectual Property Organisation. 1988. *Report Adopted by the committee of experts on biotechnological inventions and industrial property*, BIOT/CE/IV/4. Geneva: WIPO, 16.

83   WIPO, Revised Report BIOT/CE/III/2, above n 76, 31, 69.

84   The International Federation of Industrial Property Attorneys maintained that if a plant has been modified at the cellular level, then all material consisting of patented plant cells should fall with the scope of claims to the plant cells, including fruits, flowers,

presented in the present tense, indicating that it was envisaged that the properties of the protected product were being actively exploited at the time of infringement, or were at least capable of being exploited in the future.

However, no consensus was reached on this proposal and, at its third meeting in June/July 1987, the Committee recommended that Suggested Solution 13 be revised. Delegates adopted two opposing positions. On the one hand, some delegations were prepared to endorse Suggested Solution 13 only if the scope of protection was narrowed; on the other hand, a number of delegations were of the view that the scope of protection was unduly limiting – in particular, the proviso that the genetic information be of *essential* importance for the industrial applicability of the second product.[85]

In response, the International Bureau attempted to clarify the wording of the proviso. In doing so, it followed the recommendation of one unidentified delegation to replace the reference to the 'industrial applicability of the protected genetic information' with the requirement that the genetic information must be expressed in the relevant matter:

> Patent protection for a product that consists of, or contains, particular genetic information as an essential feature of the invention shall extend to any matter containing the patented product or obtained from the patented product provided that the said genetic information is contained and expressed in the said matter.[86]

Here again, the similarities between the revised Suggested Solution (now Suggested Solution 9) and the second incarnation of Art 9 (Art 12 of the Amended Original Proposal) are obvious. Notably, the reference to a 'second product' has been replaced with 'any matter'; however, the active participle ('expressed') is retained. The composite expression, 'contained and expressed', also confirms that the mere presence of the genetic information in the product in which it has been incorporated is insufficient to establish infringement.

This revised proposal proved even less satisfactory to the Committee. One group of delegates recommended that the Suggested Solution be deleted, preferring to have the question of infringement determined according to 'ordinary principles of patent law'.[87] If inventors wished to obtain protection for microorganisms, then

---

seeds, cuttings, cell lines, modified cells, as well as the genes with which the cells were transformed and plasmids in which the genes are incorporated: ibid., 50.

85   World Intellectual Property Organisation. 1987. Committee of experts on biotechnological inventions and industrial property, Third session, Geneva, June 29 to July 3, 1987: Report Prepared by the International Bureau, BIOT/CE/III/3. Geneva: WIPO, 21.

86   World Intellectual Property Organisation. 1988. Revised suggested solutions concerning industrial property protection of biotechnological inventions prepared by the International Bureau of WIPO, BIOT/CE/IV/3. Geneva: WIPO, 13.

87   World Intellectual Property Organisation. 1988. Report adopted by the committee of experts, Fourth session, Geneva October 24 to 28, 1988, BIOT/CE/IV/4. Geneva: WIPO, 21.

it was open to them to draft claims for the microorganisms themselves. Extended to plants, however, this strategy would be impractical for the vast majority of countries that refused patents for plants. Indeed, this was precisely the point of the Suggested Solution: several delegations were concerned that extending the scope of protection for genetic information in this manner may have the 'unintended effect of providing for the domination of patent protection over plant variety rights',[88] and recommended that the issue be considered further by a joint meeting of the UPOV Convention members and the WIPO members. Far from being an unintended effect, this was precisely the object sought by those dissatisfied with the UPOV Convention settlement, a prominent theme in discussions about patent protection for biotechnology in the 1980s. Other delegations resisted this move, pointing out that the Suggested Solution was concerned not only with the scope of protection for genetic information introduced into plant varieties but with many other products besides.

Rehearsing arguments later made by Monsanto, a number of delegates also expressed concerns about the words 'expressed in'. They argued that, in certain cases, protected genetic information may have an effect on the physiology of the final product, without being expressed in that product.[89] To address this issue, a further revision of the Suggested Solution was prepared and discussed at the fourth and final meeting of the Committee:

> Patent protection for a product that consists of, or contains, particular genetic information as an essential feature of the invention as defined in the claim or claims shall extend to any use of the patented product as contained in any matter provided that the said genetic information is essential for its use.[90]

This too was regarded as unsatisfactory. Several delegates objected to the idea that infringement should only lie where the protected genetic information is essential for the use of any product in which it is contained, because, on general principles, liability may be established even when the protected subject matter is not essential to the use of the product in which it is contained but is only helpful or desirable.[91] The revised formulation might therefore lead to a diminution of protection for genetic information. Several other suggestions for improving the wording of the provision were accordingly advanced, including replacing all of the words appearing after 'shall extend' with the words 'to any use of the patented product in any matter in which the said genetic information is utilized', or 'to any use of the patented product in or for the production of any matter containing the said genetic information'.[92] However, the meeting was concluded without any alternative form

---

88   Ibid.
89   Ibid.
90   Ibid.
91   Ibid., 21–2.
92   Ibid., 22.

of wording being agreed upon and the matter was referred back to the International Bureau. The Committee also agreed that a joint meeting of WIPO and UPOV Convention members should be held in order to clarify the interface between plant breeders' rights and patents. Although a further meeting of the Committee was contemplated once WIPO and UPOV Convention members had finalised their discussions, the Committee undertook no further work after this meeting.

Although the delegates were ultimately unable to agree on a formulation of Suggested Solution 13/9 that was acceptable to all parties – the International Bureau acknowledged that 'the task of finding satisfactory wording was an extremely difficult one'[93] – two matters can be extrapolated from the discussions as they stood. First, although favoured by some delegations (the International Bureau did not disclose whom), at no stage during consideration by the Committee of the various iterations of Suggested Solution 13/9 was it regarded as sufficient for patented genetic information merely to be present in a greater product: something more was necessary. As the International Bureau noted in its fourth report, the Committee acknowledged that the words 'expressed in' 'added a separate requirement to the requirement that the genetic information be contained in a particular matter'.[94] Settling an appropriate form of wording that was capable of capturing this elusive, additional element escaped the grasp of the International Bureau and the Committee. However, common to each of the attempts made in this direction was the assumption that the defendant, by using the complex product, directly or indirectly, uses the patented genetic information.[95] In the first iteration of Suggested Solution 13, this is captured by the unfortunately-worded phrase 'in which the said genetic information is of essential importance *for the industrial applicability or utility of the said product*', and in the later expressions 'provided that the said genetic information is contained *and expressed* in the said matter', and 'provided that the said genetic information is essential *for its* [i.e. the greater product's] *use*'. The point conveyed by each version is that for infringement to arise, use of the complex product must of necessity entail use of the patented genetic information. As the International Bureau confirms in its questionnaire, the issue addressed by question 2.1.3 is 'whether the presence of a protected product in a more complex material (where its properties *are used* and of essential importance for the value of the complex material) means that the complex product is covered by the claim'.

---

93   Ibid., 20.

94   Ibid., 21.

95   That this was apparent from the outset can be seen from the commentary accompanying question 2.1.3 of the questionnaire addressed to governments and intergovernmental organisations, which states that the essence of the question, in technologically neutral terms, is whether the presence of a protected product in a more complex material means that the complex material is covered by the claim, 'where its properties are used and are of importance for the value of the complex material': WIPO, Questionnaire BIOT/Q1, above n 75, Chapter II, 42.

Second, the genetic information must itself be an essential feature of the invention, and there must be some direct relationship between the genetic information and the industrial applicability of the greater product in which it is incorporated (although this feature is absent from the first revised Suggested Solution, which became Suggested Solution 9). Along with the assumption that use of the complex material necessarily involves use of the patented product, the extension of protection to the greater product itself is justified by the contribution of the patented product to the utility of the greater product.[96]

The Negotiating History of Art 9 of the Directive

This was the *status quo*, so far as international efforts are concerned, at the time when European lawmakers first contemplated the introduction of a harmonised approach to the protection of biotechnological inventions. The Commission was an observer at the Third and Fourth meetings of the WIPO Committee of Experts and would therefore have been aware of the various positions adopted by countries and organisations represented at those meetings. Indeed, the initial version of Art 9, which appears as Art 13 of the Original Proposal, reproduces *verbatim* the wording of Suggested Solution 13. Likewise, Art 10 of the Amended Original Proposal reproduces the wording of the first revision of Suggested Solution 13 (Suggested Solution 9). Given the criticisms directed at these formulations by the Committee of Experts, it was almost inevitable that these proposals would be just as unacceptable to European lawmakers as they were in that forum, and so they proved to be.

The issue *du jour* when the Original Proposal for a Directive appeared in the late 1980s was the distinction between discovery and invention and whether or not biological materials, including animate objects, were patentable subject matter. Along with plants and animals, genes and DNA were the most conspicuous subjects of this debate. The Original Proposal did not contain, as the Directive does, provisions specifically addressing the patentability of genetic information, although the Commission clearly contemplated that it is, or should be, patentable. As the Explanatory Memorandum to the Original Proposal states, '[a]s a result of the Directive greater possibilities will exist for patenting products consisting of or containing genetic information, such as a particular

---

96   AIPPI supported this extension of protection '*on the assumption* that the material patented contributes significantly to the properties or characteristics of the plant or animal'. The International Group of National Associations of Agrochemical Products (GIFAP) and the Pacific Intellectual Property Association (PIPA) commended the extension on the basis that 'the commercial value of an invention resides very often in the plant or animal.' Without such an extension of protection 'no meaningful protection could be obtained ... and damages in cases of infringement would be difficult to assess'. World Intellectual Property Organisation. 1988. *Industrial Property Protection of Biotechnological Inventions: Newly Revised Report prepared by the International Bureau*, BIOT/CE/IV/2. Geneva: WIPO, 49.

DNA segment'.[97] Support for this presumption was found in Arts 2 and 8, which then provided that an invention does not cease to be patentable merely because it is living or occurs naturally, respectively.

However, the scope of protection conferred upon claims to genetic information remained uncertain, particularly 'for those inventions which do not permit their direct exploitation but which must become part of another entity in order to be used effectively'.[98] DNA, as the material substrate for genetic information, is a primary example of such subject matter. As American historian of science, Evelyn Fox Keller, recently remarked:

> By themselves ... sequences of nucleotides [i.e. that code for a protein] don't do anything: DNA is an inert molecule ... By themselves, the entities we call genes do not act; they do not have agency. Strictly speaking, the very notion of a gene as an autonomous element, as an entity that exists in its own right, is a fiction. In order for a sequence of nucleotides to become what is conventionally called a gene requires that the sequence be embedded in a cellular complex that not only reads, translates, and interprets that sequence, but also defines it, giving it its very meaning.[99]

Given the variety of different views regarding the proper scope of protection for inventions of this type, 'for which existing patent laws provide no solution',[100] the Commission insisted that patent rights 'must be legislatively mandated' for 'any

---

97   Commission of the European Communities. 1989. Proposal for a Council directive on the legal protection of biotechnological inventions, COM(88) 496 final SYN 159. *Official Journal of the European Communities*, 13.1.89, C 10/3.

98   Ibid.

99   Keller, E. 2010. *The Mirage of a Space between Nature and Nurture*. Durham: Duke University Press, 6.

100   This view was shared by many contemporary commentators. For example, Andrew Christie argued that '[w]here a patented gene product consists of or contains plant genetic information (for example, a new gene), there is some uncertainty as to whether incorporation of the genetic information into another plant constitutes an infringement of patent. *Resolution of this issue requires a policy decision.* A suitable approach is to regard the incorporation as an infringement if the genetic information is of commercial significance to the other plant' (emphasis supplied): Christie, A. 1989. Patents for plant innovation. *European Intellectual Property Review*, 11(11), 394–408, 408. Writing shortly after the rejection of the Original Proposal in 1995, Joseph Straus lamented that 'Art 10 [the amended first proposal's precursor to Art 9] *for the first time* would have clearly marked out the scope of protection of patents in genes ... the Directive would have secured throughout the member States not only the patentability of DNA ... but would also have harmonized the effects of patents granted on genes': Straus, J. 1995. Patenting human genes in Europe – Past developments and prospects for the future. *International Review of Industrial Property and Copyright Law*, 26(6), 920–50, 944. See also Roberts, T. 1995. Draft directive on legal protection of biotechnological inventions. *European Intellectual Property Review*, 17(4), D116–17.

final product whose utility, commercial value or industrial applicability depends on a patented invention'.[101] Unless this was done, there would be:

> insufficient incentive for ensuring that necessary research is undertaken to accord patent protection only to material which on its own has no commercial value ... Without Art 13, it might be considered that the patent protection of a biological product would be lost if such product becomes part of a more complex final product even though such biological product is of essential importance for commercialising the final product.[102]

Here again we see emphasised the nexus between the patented genetic information and the product in which it is incorporated: the commercial value and industrial applicability of the latter is indebted to the former, which in turn is incapable of being exploited unless it is incorporated into the latter. Intrinsic to this co-dependent relationship is the notion that the defendant, by using the greater product, directly or indirectly makes use of the genetic information.

Against this extension of protection three objections were raised when the Original Proposal reached the European Parliament. First, extending the scope of patent protection for genetic information in this manner raised the prospect that claims to isolated 'human' DNA sequences might encompass the gene in its natural environment, i.e. the human body. Although this is an unlikely and unintended scenario, the Original Proposal's failure clearly to disclaim the patentability of human beings was widely criticised.[103] Article 3(a) of the Amended Original Proposal rectified this omission. However, the derogation presently contained in Art 9 (the reference to Art 5(1)) was not inserted until the Second Proposal was introduced in December 1995 (although this does not appear to have been a factor in the final rejection of the Original Proposal).

The second criticism was one of proportionality: is it justifiable to extend protection to any material in which the protected genetic information has been incorporated, of which the genetic information, in both a quantitative and qualitative sense, is but a minor component? This was considered by some as inappropriate because 'manufacturing steps were required to obtain the variety from the

---

101 See also Crespi, R. 1992. Patents and plant variety rights: Is there an interface problem? *International Review of Industrial Property and Copyright Law*, 23(2), 168–84, 181–2: 'In the plant gene context the patent must ... extend to the DNA *in the plant* and to this extent is exercisable in relation to the whole plant ... [because] [t]he commercial product in which this value-adding DNA component makes its presence felt is the whole plant ... the effect of the part is commensurate with the performance of the whole.'

102 Commission of the European Communities, above n 97, C 10/52.

103 See Economic and Social Committee. 1989. Opinion on the proposal for a council directive on the legal protection of biotechnological inventions. *Official Journal of the European Communities* C 159/112, [2.15].

patented product.'[104] According to one expert on biotechnology patents who was consulted by the Commission, Stephen Crespi, the extension of protection provided by Art 9 is justifiable on the basis that 'the effect of the [genetic information] is commensurate with the performance of the whole.'[105] The situation is thus distinct from that of a mechanical invention, such as a carburettor or rivets used in the construction of a ship, where a claim to the car or ship will not ordinarily be justified having regard to the technical contribution of the part in relation to the whole.[106] Whilst a motor vehicle would not be operative without a carburettor, nor a ship seaworthy without rivets, these are but means to an end; for GM plants, the genetic information is the *raison d'être* of the variety, even though it recruits biological processes to produce its intended function. As the Commission observes in the Explanatory Memorandum accompanying the Original Proposal, this step is only justified where 'the particular industrial applicability or usefulness of a variety directly results from an invention which has been patented' in the sense that the patented invention is 'of essential importance for the utility or industrial applicability of the final product'.[107]

Of course, an inventor may always prepare claims for the products in which the genetic information is incorporated, and ordinarily will do so. However, at least so far as plant and animal varieties are concerned, at the time when the Original Proposal was first prepared there was no certainty that such claims could be obtained, even where the invention was capable of being implemented in more than one variety.[108]

The third criticism of Art 13 ventriloquised the debate within the Committee of Experts regarding the effect of claims to DNA sequences that are incorporated into plant varieties. Here it was argued that extending the scope of protection for such claims in the manner envisaged by Art 13 would amount to *de facto* protection of the plant itself, contrary to the prohibition on the patenting of plant varieties found in Art 53(b) of the European Patent Convention. The Economic and Social Committee (ESC) was especially critical of the fact this 'enlargement of the field of patentability' had been made without proper consultation with all interested parties and was largely based on 'consultations representing industry's interests only'.[109] The ESC was also critical of the Commission's approach of 'dealing first with patentability and only after with how this would affect plant breeders'

---

104 Commission of the European Communities, above n 97, C 10/ 51.

105 Crespi, above n 101, 182.

106 Ibid.

107 Commission of the European Communities, above n 97, C 10/ 51. However, see Pottage, A. 2007. The socio-legal implications of the new biotechnologies. *Annual Review of Law and Social Science*, 3, 321–44, 327–8.

108 T 49/83 *Ciba-Geigy/Propagating Material* [1979–1985] E.P.O.R. 758; T 320/87 *Lubrizol/Hybrid Plants* [1990] E.P.O.R. 173; T 356/93 *Plant Genetic Systems/Glutamine Synthetase Inhibitors* [1995] E.P.O.R. 357.

109 Economic and Social Committee, above n 103, C 159/11.

rights'. In the ESC's view, a 'dual approach' to these issues 'would have led to a better balance between breeders' rights and patents, which preserved the rights and interests of the parties concerned (farmers, agricultural cooperatives, breeders, researchers, industry)'.[110] It accordingly recommended that Art 13 be deleted.[111]

The ESC was principally concerned with preserving the 'farmer's privilege' afforded under the UPOV Convention and feared that this right would be usurped unless the Directive was amended to incorporate an equivalent exemption in respect of patent rights. This soon emerged as 'the key issue' within Parliament.[112] Despite having the support of the 'vast majority' of the members of Parliament, the Commission was initially dismissive of this suggestion 'for legal, technical and economic reasons,' and only relinquished this position once it realised that its unwillingness to accommodate Parliament's demand might jeopardise the Directive's progress. Given the 'vigour' with which Parliament expressed this demand, this was a 'political sign' that the Commission felt unable to ignore.

When an Amended Proposal was re-introduced in mid-December 1995, it contained a new Art 13 incorporating Parliament's demand, although the Commission was at pains to emphasise that the exception was not in the nature of a 'privilege' so much as a derogation from the extent of protection conferred by a patent.[113] However, the Commission was unwilling to accede to the ESC's demand that Art 13 be deleted. Rehearsing the argument that later found favour with the Enlarged Board of Appeal of the European Patent Office in *Novartis/ Transgenic Plant*,[114] the Commission explained that 'an exclusion of varieties is not synonymous with being free from the scope of a relevant patent', especially where the invention 'concerns a generic concept which is characterised by new generic information ... which can be realised in a multitude of different varieties.'[115]

The Amended Proposal also reintroduced in 'corrected form' (as the Explanatory Memorandum put it) Art 13, now Art 11. The correction mirrored the amendment made to Suggested Solution 13 following the third meeting of the Committee of Experts in October 1987 which, in substance, replaced the expression 'and is of essential importance for its industrial applicability or utility' with 'and in which the genetic information is incorporated and expressed'. The requirement that the

---

110   Ibid.

111   'Patent law cannot apply to plants and/or animals in which the invention has been incorporated, in accordance with the reworded Art 3': ibid., C 159/12. The ESC's amendment to Art 3 provided that '[m]icro-organisms and the genetic components of plants and animals up to protoplasms shall be considered patentable subject matter'.

112   Commission of the European Communities. 1992. *Amended Proposal for a Council Directive on the Legal Protection of Biotechnological Inventions*, COM(92) 589 final – SYN 159, 16.

113   Ibid., 17.

114   [1999] E.P.O.R. 123.

115   Commission of the European Communities, *Proposal for a Directive on the Legal Protection of Biotechnological Inventions*, COM(88) 496 final – SYN 159, 17 October, 1988, 52.

genetic information must be an 'essential characteristic of the invention' was also deleted. The Explanatory Memorandum provides no further guidance as to the reasons for this amendment, other than to clarify that the extension of protection applies to all material in which the protected genetic information is incorporated 'irrespective of its parentage, which is dealt with in Art 10 [Art 8 of the Directive]'.[116]

History, of course, records that the Amended Proposal was rejected by Parliament on 1 March 1995.[117] When a Second Proposal was re-introduced by the Commission later that year,[118] it retained the wording of Art 11 of the Amended Original Proposal. By subjecting the operation of Art 11 to Art 3 (the precursor to Art 5 of the Directive), which at the time provided that 'the human body and its elements in their natural state shall not be considered patentable inventions,' the Second Proposal for the first time also made clear that the scope of protection for genetic information does not extend to the human body, thereby addressing a persistent concern with the earlier proposals.[119] Other than clarifying that the concept of genetic information automatically incorporates the material substratum on which it is based (such as DNA),[120] the Explanatory Memorandum remained silent regarding the intended application of Art 11.

Parliament subsequently moved 66 amendments to the Second Proposal which were addressed by the Commission's Second Amended Proposal published at the end of August 1997.[121] So far as Art 11 (now Art 10) is concerned, only two of

---

116 Commission of the European Communities, above n 112, 15.

117 European Parliament. 1995. *Decision on the Joint Text Approved by the Conciliation Committee for a European Parliament and Council Directive on the Legal Protection of Biotechnological Inventions*, C4–0042/95 – 94/0159(COD), Co-decision Procedure: Third Reading. *Official Journal of the European Communities* C 68/26.

118 The Proposal was submitted to Parliament on 25 January 1996.

119 Commission of the European Communities. 1996. *Proposal for a European Parliament and Council Directive on the Legal Protection of Biotechnological Inventions*, COM(95) 661 final–95/0350(COD), 14, 16–17. *Official Journal of the European Communities* C 296, 4. According to the Economic and Social Committee this was 'the most controversial point of the previous Directive'. See Commission of the European Communities. 1996. Opinion of the Economic and Social Committee on the 'Proposal for a European Parliament and Council Directive on the legal protection of biotechnological inventions'. *Official Journal of the European Communities*, 7.10.96, C 295/11, [4.3]. The Committee on the Environment, Public Health and Consumer Protection recommended that Art 11 of the Second Proposal be deleted, apparently in deference to ethical and moral concerns regarding the patenting of genes, particularly those of human origin: Commission of the European Communities. 1996. Opinion of the Committee on the Environment, Public Health and Consumer Protection on the 'Proposal for a European Parliament and Council Directive on the Legal Protection of Biotechnological Inventions', in *Committee on Legal Affairs and Citizens' Rights Report on the Proposal for a European Parliament and Council Directive on the Legal Protection of Biotechnological Inventions*, A4–0222/97, 25 June 1997, 53–61.

120 COM(95) 661 final – 95/0350(COD), ibid., 19.

121 Commission of the European Communities. 1997. Proposal for an European Parliament and Council Directive on the Legal Protection of Biotechnological

the suggested amendments were germane, amendments 47 and 58. The former proposed the insertion of a new Art 2a. Paragraph 1(a) of that provision stipulating that plant and animal varieties shall not be patentable. Amendment 58 amended Art 11 by providing that, along with human beings, the scope of protection for genetic information does not extend to plant and animal varieties in which the genetic information is contained and expressed.

The suggested amendments would therefore have rendered nugatory the very premise on which Art 9 was originally based. Unsurprisingly, this amendment was rejected by the Commission as being both 'technically and legally incomprehensible', as well as contrary to current practice under patent law.[122] By the time the Council of Ministers and the Commission published their Common Position at the end of February 1998, Art 9 had acquired its present form. The expression 'contained and expressed' was jettisoned in favour of 'contained and performs its function'.[123] The Statement of Reasons accompanying the Common Position does not offer any assistance regarding the reasons for this change in wording; in fact the Council 'thought it was more appropriate to use the words "in which the genetic information is contained and expressed"', rather than the alternate wording that was ultimately adopted.[124] Nor are they addressed by the Commission's communication of the Common Position to Parliament on 4 March 1998[125] which, other than confirming that the Common Proposal 'incorporates and completes as appropriate the amendments accepted at first reading [of Parliament]', provides only that the Common Position includes a number of changes 'which constitute technical improvements to the amended proposal'.[126] Parliament accepted the Common Position without amendment on 12 May 1998.[127]

Inventions, COM(97) 446 final – 95/0350(COD). *Official Journal of the European Communities* C 311/12.

122   Commission of the European Communities 1997. *Proposal for a European Parliament and Council Directive on the Legal Protection of Biotechnological Inventions,* COM(97) 446 final – 95/0350(COD), 5.

123   European Parliament. 1998. Common Position (EC) No. 19/98 adopted by the Council on 26 February 1998 with a view to adopting Directive 98/ ... /EC of the European Parliament and of the Council on the Legal Protection of Biotechnological Inventions (98/110/02). Official Journal of the European Communities C 110/17, 23.

124   Ibid., 31.

125   Commission of the European Communities, Communication from the Commission to the European Parliament pursuant to the second paragraph of article 189b(2) of the EC-Treaty: Common position adopted by the Council on 26 February 1998 on the proposal for a European Parliament and Council Directive on the legal protection of biotechnological inventions, SEC(1998) 360 final – 95/0350(COD). Amendments 47 and 58 are said to be addressed by Art 4 and Art 9 respectively.

126   Ibid., 8.

127   European Parliament. 1998. Decision on the common position adopted by the Council with a view to adopting a European Parliament and Council Directive on the

Given the criticisms to which the expression 'is contained and expressed' was subjected at the fourth meeting of the Committee of Experts in 1988, it is surprising that the Commission persisted with this wording until early 1998. Although the reasons for abandoning the expression are unclear, it seems likely that the decision was informed by these criticisms, along with a realisation that genetic information can be deployed not only as a transformant, but in a number of other contexts besides, such as polymerase chain reaction (PCR) and diagnostic testing, where the language of 'expression' is inappropriate.[128] That is, the words 'performs its function' may have been preferred because they capture a wider range of contexts, not simply those involving expression of genetic information in a biological system.

Whatever the reasons for that change, the negotiating history of the Directive discloses that Art 9 is intended to address at least three different, and to some extent overlapping, issues. First, the Commission was committed to ensuring that the scope of protection for claims to genetic information extended to plant varieties in which such material is incorporated, particularly where the invention involves a 'generic concept' – i.e. the invention (genetic information) can be implemented in a number of different varieties. At the time Directive was being negotiated, and at its conclusion, the question was whether or not an invention which included multiple varieties within its scope remained at large. By confirming that the scope of protection for patented genetic information includes plant varieties containing that information (provided that it performs its function in any variety containing it), Art 9 offered a solution to this conundrum.[129] Such resistance as was encountered by Art 9 in its various iterations primarily involved this issue.

---

legal protection of biotechnological inventions (C4–0132/98 95/0350(COD))(Codecision procedure: Second reading). Official Journal of the European Communities C 167/26.

128   Another reason may have been to standardise the language used throughout the Directive, which makes repeated reference to 'the function' of DNA and genetic information.

129   Ironically, a minority of the Supreme Court of Canada in *Harvard College v. Canada (Commissioner of Patents)* [2004] 4 S.C.R. 45 rejected the argument that claims to a patented oncogene and/or cells transformed with this gene would provide the inventors with sufficient protection (since any use of this gene or cell would constitute infringement), only to resurrect this argument against the defendant in *Monsanto Canada Inc. v. Schmeiser* [2004] 1 SCR 902, 904: 'Case law shows that infringement is established where a defendant's commercial or business activity involving a thing of which a patented part is a component *necessarily involves use of the patented part*' (emphasis supplied). According to the minority in *Harvard*, such claims would provide insufficient protection because 'it would be easy for "free riders" to circumvent the protection sought to be given to the inventor by the *Patent Act* simply by acquiring an oncomouse and breeding it to as many wild mice as desired and selling the offspring (probably half of which will be oncomice) to the public': [2004] 4 S.C.R. 45, [98]. However, the minority then refers to difficulties involved in enforcing process patents, which suggests some confusion regarding the effect of product patents. Judging by the result, it seems to have reconciled this confusion by the time *Schmeiser* was decided.

Second, from the outset the mere presence of protected genetic information in a complex product was regarded as insufficient to establish infringement. Each of the various iterations of Art 9 attempt to capture, with varying degrees of success, the idea that the relationship between the protected genetic information and the greater product is such that use of the greater product necessarily involves use of the protected genetic information. Hence by its use of the expression 'shall extend to any products in which the genetic material is incorporated and is of essential importance for its industrial applicability or utility', the proviso contained in the original version of Art 9 (Art 13 of the Original Proposal) directly linked the protected genetic information to the industrial applicability or utility of the product.[130] So too, underlying the use of the performative expression 'all material in which the product is incorporated and in which the genetic information is contained and expressed/performs its function' is a presumption that the user of, say, a plant variety or micro-organism which incorporates patented genetic information obtains the benefit of this information whenever it is expressed or performs its function (or remains capable of being expressed or performing its function) within the organism. Only when this relationship exists did the Commission regard the extension of the scope of protection for genetic information to materials in which it is actualised as being warranted.[131]

Finally, the Explanatory Memorandum accompanying the Original Proposal clearly contemplates that protected genetic material must be of essential importance for the utility or industrial applicability of the *final product*. In other words, if the protected genetic information is inutile within the final product, protection ceases. Although the final version of Art 9 contains no reference to final products, the phrase 'any material' unquestionably captures such products.

---

130   'Article 13 is necessary so that this important principle of patent law is explicitly recognised for inventions which do not permit their direct exploitation but which must become part of another entity *in order to be used* effectively … Without Art 13, it might be considered that the patent protection of a biological product would be lost if it becomes part of a more complex final product': COM(88) 496 final – SYN 159, above n 115, 52 (emphasis supplied).

131   '[I]f the particular industrial applicability or usefulness of a variety directly results from an invention which has been patented, then such a variety owes its unique characteristics to the effects of the invention and should therefore come within the scope of protection accorded by the patent. Where the invention is of no commercial importance for the variety, *then a different issue would be raised*. This situation is not addressed in the Directive because Art 13 specifically stipulates that the patented invention must be of essential importance for the utility or industrial applicability of the final product': ibid., 51 (emphasis supplied). The italicised passage confirms that the mere presence of protected genetic information in the complex product is insufficient to establish infringement.

Article 9 and the Principle of Absolute Protection

A number of commentators have offered a different interpretation of Art 9, relying principally on explanatory memoranda accompanying both the Original Proposal and domestic implementing legislation. Perhaps the most prominent critic of the CJEU's decision has been Michael Kock, the Global Head of I.P. (Seeds and Biotechnology) for the Swiss agricultural biotechnology firm Syngenta International AG. In a series of articles,[132] Kock has challenged the reasoning employed by both the Advocate General and the CJEU in reaching their decisions. In somewhat selective fashion, Kock emphasises various passages of the Explanatory Memorandum to the Original Proposal which allegedly suggest that the intention of Art 9 was to '"extend" the protection for DNA sequences beyond what a patent normally confers' and to 'create "greater possibilities"' for patenting products consisting of or containing genetic information.[133] According to Kock, '[i]f national patent law were applied to the current case, the finding of infringement would be an almost certain outcome.'[134] Because, on the CJEU's interpretation of Art 9, the importation of soy meal containing the DNA sequence claimed by Monsanto did not amount to infringement of that claim, Kock argues that the decision is inconsistent with legislative intent.

Clearly, Kock's object is to preserve the principle of absolute protection. Central to his critique is the claim that if the matter had been decided according to national law, rather than the Directive, a finding of infringement would almost invariably have followed because DNA, when defined, as it was in the relevant claim, solely by reference to its chemical composition and without reference to the method(s) by which it is prepared or exploited, is, like any other kind of chemical substance, subject to the principle of absolute protection.[135] On this interpretation, the importation of soy meal containing the patented DNA sequence constitutes infringement of the claim to the DNA sequence as such, because the right to import the patented DNA sequence is one of the exclusive rights reserved to the patentee under Dutch law.

---

132   Kock, M. 2010a. Purpose-bound protection for DNA sequences: In through the back door? *Journal of Intellectual Property Law and Practice*, 5(7), 495–513; Kock, M. 2010b. Patent protection for DNA sequences – To be or not to be? *Journal of Intellectual Property Law and Practice*, 5(11), 754–56; Kock, M. 2011. Court of Justice of the European Union limits patents on DNA sequences: Much ado about nothing or the beginning of erosion for biotech patents. *Bio-Science Law Review*, 11(1), 3–112.

133   Kock 2010a, ibid., 505.

134   Kock 2011, above n 132, 112. According to Kock, 'all Member States, with the exception of France for human sequences, prior to the Biopatent Directive were and, up to now, are providing absolute protection for all patentable subject-matter, including DNA sequences': Kock 2010a, ibid.

135   See also Committee on Legal Affairs and Citizens' Rights. 1997. *Report on the Proposal for a European Parliament and Council Directive on the Legal Protection of Biotechnological Inventions*, 37. [Online]. Available at: http://www.europarl.europa.eu/sides/getDoc.do?type=REPORT&mode=XML&reference=A4–1997–222&language=EN [accessed 14 March 2014]: 'Protection of substances is absolute'.

However, it is far from self-evident that the application of this principle to the claims in suit would inevitably result in a finding of infringement. In the cognate English proceedings,[136] which were decided according to general patent law principles rather than according to the Directive,[137] the late Justice Pumfrey held that importation of soy meal containing the claimed DNA sequence did not infringe those claims, nor the patent's process claims.[138] In contrast to the litigation in the CJEU, which was principally concerned with statutory construction, the English litigation was largely decided by claim construction. Two features of the way in which the patent was drafted were decisive. The first involved the inclusion of the term 'isolated' in each of the DNA sequence claims. Although commonplace in US patents, where it is employed to confer novelty upon claims to naturally occurring materials, Justice Pumfrey noted that "isolated' was a surprising word to use considering that the traces of DNA sequences contained in the allegedly infringing soy meal were present in admixture with other materials.[139]

Secondly, Monsanto's allegation of infringement was difficult to reconcile with the structure of the specification itself. 'What is striking about these claims', Justice Pumfrey said, 'is that down to claim 14 (the method of transforming the

---

136   *Monsanto Technology LLC v. Cargill International SA* [2008] F.S.R. 153.

137   The United Kingdom is the only jurisdiction to apply a transitional period to the introduction of the Directive, the validity of which must be questionable given the CJEU's answer to question 3. However, as Michael Fysh QC has observed, to have decided the case under the Directive would have contravened the common law presumption against giving legislation retrospective effect, without clear language to the contrary in the offending statute. Fysh, M. 2011. *IP Protection and Enforcement Strategies*. Unpublished conference paper, UNION Congress, Brussels, 21–24 June 2011.

138   Justice Pumfrey held that the process claims were not infringed because the DNA sequences were not 'directly obtained' by the patented process, an appellation reserved for the 'immediate products of the process' or, where the patented process is an intermediate stage in the manufacture of some ultimate product, that product, but only if the product of the intermediate process still retains its identity'. Here, the direct product of the claimed process was original transformed plant produced in accordance with that process. Moreover, 'product' is defined in material terms: it must be possible to trace the starting materials used in the process into the product that emerges from the process. This product may be subject to further transformation, but it must retain its material identity. 'It must be close to the truth,' said Justice Pumfrey, 'that the generation of plants producing the beans from which the ... meal was manufactured did not have an atom in common with the original transformed plant'. Justice Pumfrey also cautioned against talking of reproductive material 'having in some way passed between the generations'. Whilst reproductive material does pass between the first and second generations, 'the same material does not pass further. Copies pass thereafter.' In other words, there must be physical continuity between the starting materials acted upon by the claimed process, and the products that come into being at the end of this process; a genealogical connection will not suffice: see *Monsanto Technology LLC v. Cargill International SA* [2008] F.S.R. 153, 173–4.

139   See also Krauss and Takenaka, above n 68, 199–200.

plant) they all essentially relate to the laboratory work.'[140] 'It makes sense in this context,' his Honour continued, 'to have claims relating to each of the stages in transforming the plant, starting with the isolated sequence and proceeding through the sequence appropriately topped and tailed to transforming a plant using that molecule.'[141] Given this context, Justice Pumfrey concluded that 'isolated' means 'separated from other molecular species in the form of a purified DNA fragment' for the purpose of cloning and amplifying in a plasmid DNA;[142] that is, 'the DNA is ready for use within a laboratory to carry out a recombinant DNA technique.'[143] Accordingly, the allegation of infringement of these claims failed, notwithstanding the fact that the DNA sequence was found to be present in the allegedly infringing product.[144]

Whilst claim construction was decisive in the English proceedings, case law in other jurisdictions confirms that a finding of infringement on the same facts would not be a *fait accompli* under national law. In its Grasherbizid decision,[145] the District Court of Düsseldorf held that claims to chemical intermediates (trifluromethylpyridyl(oxy/thio) phenols) used in the production of a commercial herbicide, Fusilade, were not infringed by the use of that herbicide, despite the fact that a residual amount of the intermediates (0.1–0.4 per cent) was present in the herbicide as incomplete reaction products. Whilst intermediate products enjoy absolute protection under German law, the Court held that the scope of product claims to a chemical intermediate does not extend to the final product in which the intermediate substance is no longer present as such but has, by chemical

---

140   *Monsanto Technology LLC v. Cargill International SA* [2008] F.S.R. 153, 183.

141   Ibid.

142   Ibid.

143   Cohen, S. and Morgan, G. 2008. *Monsanto Technology LLC v. Cargill*: A matter of construction. *Nature Biotechnology*, 26(3), 289, 290. The same interpretation was arrived at by The Hague District Court, which held that the defendants 'correctly take the view that these claims cannot have been infringed because the DNA is not present in isolation, but is contained in the soya flour ... The average person skilled in the art will understand the concept of an isolated DNA in the sense it is ordinarily understood, i.e. a DNA that normally is released from the cell of an organism for further processing': *Monsanto Technology v. Cefetra*, above n 34, [4.4].

144   Monsanto's action failed for the further reason that they did not submit sufficient evidence to establish that the DNA sequence found in the meal conformed to the other essential feature of the product claims – that the expression product of the DNA sequence is a Class II EPSPS enzyme. Monsanto's appeal against Justice Pumfrey's decision was dismissed by the Court of Appeal, which considered that his Honour was justified in the construction at which he arrived.

145   *'Grasherbizid'* G.R.U.R. 1987, 896. See also *Synaptic Pharmaceutical Corp. v. MDS Panlabs, Inc.* 265 F. Supp. 2d. 452 (2002), a decision of the District Court of New Jersey which seems to imply, without finally deciding, that a claim to an isolated DNA molecule does not cover recombinant cells genetically-engineered by the introduction of a plasmid containing the claimed DNA.

transformation, become fully absorbed into a new substance whose physical characteristics and technical applicability differ from those of the intermediate. Moreover, where the intermediate is present in the final product in small amounts, use of the final product does not constitute infringement of the product claim to the intermediate if the intermediate makes no contribution to the functioning of the final product. As the claimed intermediates had no herbicidal activity, there was no infringement.[146]

Given that DNA is sometimes likened to a starting material or an intermediate, which is employed in a biotechnological process to produce a particular result (for example, the expression of a particular protein conferring glyphosate-herbicide resistance in plants),[147] the court's reasoning seems particularly felicitous to the Monsanto litigation. Kock attempts to downplay the relevance of this decision by characterising it as an example of a *de minimis* defence 'which has not been adapted by higher courts and cannot be seen as established law'.[148] However, the Court clearly felt that the intermediate was not being exploited in the final product and the better, and more widely accepted, view is that there was no infringement at all.[149]

Nor is the legislative intention that Kock imputes to Art 9 readily apparent. In the first place, the passages on which he relies must be read in their proper context. The Commission's statement that '[a]s a result of the Directive greater possibilities will exist for patenting products consisting of or containing genetic information, such as a particular DNA segment,' refers not to the extension of protection given by Art 9 but to the fact that, by clarifying the distinction between discovery

---

146 The Court also observed that the patentee had the opportunity to obtain patent protection for the subject matter serving as the basis of the technical merit (i.e. the intermediate and the process for the production thereof) or for the subject matter where the merits become manifest (the end product or its use). Indeed, the patentee had obtained a second patent covering both intermediate compounds and derivatives thereof (i.e. end products) that included within its scope the defendant's herbicide. However, that patent was anticipated (by one day) by the defendant's own patent for the herbicide.

147 See *Howard Florey/Relaxin* [1995] E.P.O.R. 541, 545: 'DNA is ... a chemical substance which carries genetic information and can be used as an intermediate in the production of proteins which may be medically useful.'

148 Kock 2010a, above n 132, 504.

149 Benyamini, A. 1993. *Patent Infringement in the European Community patent: Studies in Industrial Property and Copyright Law (Volume 13)*, Weinheim: VCH, 97–8; Hansen, B. and Hirsch, F. 2008. *Protecting inventions in chemistry: Commentary on chemical case law under the European Patent Convention and the German Patent Law*, Weinheim: Wiley-VCH, 345–7; Straus, J. 2008. The scope of protection conferred by European patents on transgenic plants and on methods for their production, in *Festskrift till Marianne Levin* edited by A. Engelbrekt, *et al.* Stockholm: Norstedts Juridik, 2008, 643–57, 648–9; Kamstra. G. 2011. EU legal and regulatory update, EU and Germany: Denial of absolute product protection for DNA patents (CJ Decision in *Monsanto v. Cefetra*, C–428/08) – A German Perspective. *Journal of Commercial Biotechnology*, 17, 109–10, 110.

and invention, the Directive would clearly establish that DNA sequences which have been isolated from their natural environment are patentable. Only once this threshold had been breached, could the scope of protection conferred on genetic information be considered. Moreover, the 'clear notice' given by the Commission to the Council of 'its intention to propose ... specific measures to improve patent protection of biotechnological inventions' was expressed 'in light of the negative impact which differences in national laws have on intra-Community trade and on the ability of industry to treat the common market as a single environment', and offers little succour to the argument that the Commission intended Art 9 to improve patent protection for biotechnological inventions over and above that generally applied at the national level.

Although the negotiating history is certainly ambiguous regarding the intended effect of the term 'extends' in Art 9, it is questionable whether this term subserves the purpose Kock ascribes to it. As the report of the Commission to the Council and Parliament in 2005 observes, Arts 8 and 9 might be seen as arguing for a broader scope of protection rather than a restricted one.[150] On the other hand, when the Second Proposal was debated in Parliament in 1996, the Economic and Social Committee was of the view that these Articles 'clearly circumscribe the scope of protection, limiting it to any biological material obtained through multiplication or propagation on the basis of an invention'.[151] However, it must be remembered that Art 9 originated in the WIPO Committee of Experts, where the intention was to put forward proposals for harmonising the treatment of biotechnological inventions at an international level. Given that many countries did not at that stage allow claims to plant varieties, Suggested Solution 13, which was replicated verbatim in Art 13 of the Original Directive, was developed as a means of circumventing these exclusions: by extending the scope of protection for genetic information to products in which the genetic information is incorporated, Suggested Solution 13 rendered the necessity of drafting claims to plant varieties as such otiose. The same principle informed the preparation of Art 8(2), which the International Bureau pointed out in 1987 extended protection to plant varieties produced by means of a patented process, irrespective of whether plant varieties were patented as such, 'by analogy to the widespread former practice of having the extension of process protection to products prevail over the exclusion of products *per se* from patent protection'.[152] This remained of paramount concern during the negotiation

---

150 Commission of the European Communities. 2005. Report from the Commission to the Council and Parliament: Development and Implications of Patent Law in the Field of Biotechnology and Genetic Engineering, COM(2005) 312 final – SEC(2005) 943, 3–4.

151 Economic and Social Committee. 1996. Opinion of the Economic and Social Committee on the Proposal for a European Parliament and Council Directive on the Legal Protection of Biotechnological Inventions. Official Journal of the European Communities C 295, 11, 15 ([4.8.1]).

152 WIPO. 1987. Revised Report BIOT/CE/III/2, above n 76, 29. By analogy with this former practice, which originated in Germany in the late nineteenth century, 'the

of the Directive.[153] The term 'extends' might therefore be treated as synonymous with 'includes', rather than increasing the scope of protection beyond that which prevails under national law.[154]

Article 9 and the Doctrine of Exhaustion

Perhaps the greatest difficulty confronting Kock's analysis of the legislative intention is, as the CJEU observed, that it would render the 'extension' of protection conferred by that provision otiose. By definition, the protection enjoyed by DNA sequences under national law is absolute; therefore, it is logically impossible for protection to be extended beyond what is already provided by national law. However, Kock argues that this interpretation 'ignore[s] the legislative intent of Art 9'.[155] Along with Art 8, the intention of Art 9, Kock tells us, is to 'avoid exhaustion or easy circumvention [of patent protection] by simply propagating the direct product' of a patented process for the production of self-replicable biological material, or of patented genetic information incorporated in self-replicable biological material:[156]

> It was of primary concern that patents could be simply circumvented by easy propagation of biological material after sale. This 'first-sale exhaustion' was seen as a threat to capturing value from reproducible material such as seeds.

product protection by means of extension of a particular process patent to a plant, an animal or a variety thereof should even prevail in cases where plants or animals or varieties thereof are *per se* excluded from patent protection.'

153   In relation to Art 8(2), see COM(88) 496 final – SYN 159, above n 115, 50–51. Article 12(2) provided that the extension of protection conferred by Art 12(1) (to products obtained from a patented process for the production of living matter or other matter containing genetic information and permitting its multiplication in identical or differentiated form) 'shall not be affected by any exclusion of plant or animal varieties from patentability'.

154   See Westerlund, L. and Kamstra, G. 2008. Directive 98/44/EC – Biotech Directive in *Concise European Patent Law* (second edition) edited by R. Hacon and J. Pagenberg. Alphen aan den Rijn: Kluwer Law International, 419: 'The combined effect of arts. 8 and 9 is that when material in which the product is incorporated is covered by a claim to the product, also *included* within that coverage are subsequent generations of that material if it reproduces itself.'

155   Kock 2011, above n 132, 6; Kock 2010a, above n 132, 504.

156   See also Mohan-Ram, V., Peet, R. and Vlaemminck, P. 2011. Biotech patent infringement in Europe: The 'functionality' gatekeeper. *John Marshall Review of Intellectual Property Law*, 10, 540–52, 548. Kock also argues that Art 9 of the Directive should be read as effecting a minimum, rather than absolute or exhaustive, approximation, with the result that the provision introduces a minimum standard of protection. Member states could then choose to implement this lesser form of protection, or continue to apply the principle of absolute protection to genetic information. However, this interpretation was rejected by the CJEU.

Explicit reference is also made [in the explanatory material for what later became Art 8(2)] to propagation in countries where no patent rights exist and a subsequent import in the EU. A potential infringer could piggyback on the invention by simply propagating the plant with the sequence from the seed sold by the patentee.

Exhaustion by sales of the reproducible material occurs only if the produced material is not re-used as reproductive material (for example, as food or feed product) or if the gene in consequence of the propagation has lost its functionality. The latter consideration is the reason for the requirement to *'perform its function'*.[157]

On this beguiling interpretation, there is no inconsistency between the principle of absolute protection and Art 9: along with Art 8, Art 9 simply closes 'the gap of patent exhaustion but avoids an extension to non-functional subject matter'.[158]

However, it is nonsensical to describe the patentee's exclusive right of reproduction as being exhausted in situations where patented genetic information has lost its functionality because the invention, having become inutile, is no longer capable of being worked (in the form in which it subsists). In other words, the patentee has no rights in these circumstances that are capable of being exhausted because the defendant is reproducing or using something other than the invention. The doctrine of exhaustion excuses acts that would, on a literal reading of the patent grant, otherwise amount to infringement. Put differently, what is incapable of being protected – namely, genetic information having no useful function – is incapable of being infringed.

Moreover, Kock supports this interpretation of the legislative intent by again selectively emphasising certain passages of the Explanatory Memorandum accompanying the Original Proposal that relate almost exclusively to Art 8.[159] With the exception of the passage referred to above ('greater possibilities'), which is recruited by Kock to emphasise that the intention of the Directive is to provide 'greater possibilities' for patenting products consisting of or containing genetic information, Kock is unable to identify any part of the Explanatory Memorandum or any other document produced during the negotiation of the Directive, specifically relating to the legislative intent of Art 9. Despite this, he maintains that 'Art 9 must be interpreted in this spirit,'[160] apparently for no other reason than the fact that Art 9 appears with Arts 8 and 10 in Chapter 2 of the Directive.[161]

---

157 Kock 2011, above n 132, 6.

158 Kock 2010a, above n 132, 505.

159 In his later article Kock also refers to passages of the Explanatory Memorandum to the Original Proposal dealing with an early iteration of Art 11. See Kock 2011, above n 132, 6.

160 Kock 2010a, above n 132, 505.

161 Ibid., 504, 505; Kock 2011, above n 132, 6.

Yet, the legislative history of the individual provisions included in Chapter 2 provide no sure guide to this contextual interpretation. In the Original Proposal, the doctrine of exhaustion, as applied to biotechnological inventions, was separately dealt with by Art 11 (Art 10 of the Directive). As originally drafted, Art 11 applied only to self-replicable 'products enjoying patent protection'.[162] Article 8(1) of the Directive had no equivalent in the Original Proposal, having been introduced as an amendment to the Original Proposal at the behest of Parliament. Likewise, what is now Art 8(2) was the subject of a separate provision (Art 12 of the Original Proposal), which provided an extension of protection in respect of patents for 'processes for the production of living matter or other matter containing genetic information to the products initially obtained by the patented process', *and also* to 'the identical or differentiated products of the first or subsequent generations obtained therefrom, said products being deemed also directly obtained by the patented process'.[163]

Whilst the Commission was undoubtedly concerned to ensure that the patentee's right to prevent the use of first and subsequent generations of living matter or 'other matter containing genetic information' was not exhausted by the marketing of that matter – as evidenced by the use of the conjunctive '*but also to*' in Art 12 of the Original Proposal[164] – this was not its only concern.[165] The Explanatory Memorandum makes it clear that the Commission also wished to provide vindication for extending the scope of protection for patented processes to subsequent generations of living material and not simply those immediately obtained by use of the patented process and marketed by the patentee or with their consent. The Commission considered the extension of protection to subsequent generations of products to be justified insofar as the 'properties which were initially

162   Article 11 of the Original Proposal provided: 'If a product enjoying patent protection and put on the market by the patentee or with his consent is self-replicable, the rights conferred by the national patent shall not extend to acts of multiplication and propagation only where such acts are unavoidable for commercial uses other than multiplication and propagation.'

163   Original Proposal, Art 12(1).

164   Suggested Solution 13/9 originally also provided that the scope of protection for genetic information shall extend to any product containing the patented product '*or obtained from such product*'.

165   In the Introductory Note to a questionnaire distributed by the OECD Directorate for Science, Technology and Industry, to the report on which the WIPO Committee of Experts and the drafters of the Directive are also indebted, it was observed that '[p]erhaps, it is not yet sufficiently understood that a basic, and from the point of view of patent law, significant difference, separates microbiological from all other technical innovations. This is the fact that the starting material, the micro-organism, is self-replicating under suitable conditions:' see Savers, S. 1982. International Organization Circulates Questionnaire on Whether National Patent Laws Meet the Challenge of the New Biotechnology. *Biotechnology Law Report*, 1, 58–9; 66–70, 70–71.

obtained by the process are still present and are determinative of their value'.[166] Without this extension of protection, 'the effect of the patent rights conferred by the process would be completely nullified' as it would also be if the product of the process is *per se* denied protection, as is the case with plant and animal varieties. To this end, Art 12(2) provided that the extension of protection conferred on a patented process by Art 12(1) 'shall not be affected by any exclusion of plant or animal varieties from patentability'.

Notably, neither Art 8 or Art 9 were subject to Art 11 (now Art 10 of the Directive), as they are in the Directive, although material incorporating protected genetic information was presumably covered by Art 11 to the extent that the material consisted of 'self-replicable material'. It is perhaps also of note that, in contrast to Art 8, Art 9 refers only to 'material' rather than 'biological material', which might be taken to suggest that it is not solely concerned with preventing exhaustion of rights in subsequent generations. Indeed, one of the criticisms that emerged during the discussion of the equivalent of Art 9 in the Committee of Experts was the pre-occupation of some delegations with the implications of the extension of protection for plant varieties. As was pointed out there, Suggested Solution 13 was of much wider import than its application to GM plants alone.

Parliament recommended a number of amendments to both of these Articles. First, it recommended replacing the definition of 'self-replicable matter' found in Art 19 of the Original Proposal with 'biological material', defined as 'any self-replicating living matter and any living matter capable of being replicated through a biological system'.[167] It also recommended that Art 11 be amended to read:

(1) The rights conferred by a patent the subject matter of which is a biological material with particular essential qualities shall extend to all biological materials that are obtained from the primary material by multiplication or propagation and possess the same properties.

(2) Rights within the meaning of the preceding paragraph shall not extend to biological material that is obtained from biological material that has been

---

166 COM(88) 496 final – SYN 159, above n 115, 50. The extension of protection to identical or differentiated products of the first or subsequent generations was also necessary to ensure that the patent could not be evaded by working the process in a country where the process is not patented, and a different product from that which directly results from the use of the process (for example, a plant regenerated from a patented seed or cell) is then imported into a Member State.

167 European Parliament. 1992. Proposal for a Council directive COM(88) 0496 – C3–0036/89 – SYN 159: Proposal for a Council directive on the legal protection of biotechnological inventions. Official Journal of the European Communities C 305, 160, 163 (Amendment No. 14). Article 19 of the Original Proposal defined 'self-replicating material' as including 'the genetic material necessary to direct its own replication via a host organism or in any indirect way, e.g. as comprising, inter alia, seeds, plasmids, DNA sequences, protoplasts, replicons and tissue cultures'.

marketed by the patent holder or with the latter's agreement to the extent that such multiplication or propagation result from the application for which the material concerned had been marketed.[168]

Parliament recommended that Art 12(1) be amended in the same way, so that 'processes for the production of living matter or other matter containing genetic information' was replaced with 'a process that enables the production of biological material with specific essential characteristics', with the proviso that the products derived from the use of the process, or derived through multiplication or propagation, possess these same properties. It also recommended that the limited form of exhaustion effected by Art 11(2) be applied *mutatis mutandis* to the extension of protection provided by Art 12(1).[169]

When the Commission presented the Amended Original Proposal, it merged these recommendations relating to Arts 11(1) and 12(1) into a single provision, Art 10, and applied the limited form of exhaustion provided by Art 11(2) to both the 'biological material' and processes referred to in Art 10. Subject to a number of further amendments referred to in the Common Position of the Commission and the Council, these provisions remained essentially in the same form and structure when the Directive was finalised.

Unfortunately, the Economic and Social Committee was the only parliamentary Committee whose report was published. Thus, the reasons for the introduction of Art 11(1) (Art 8(1) of the Directive) remain obscure. Moreover, it has no equivalent among the Suggested Solutions produced and debated by the WIPO Committee of Experts. However, it is quite clear that the Committee was not exclusively concerned with the application of the doctrine of exhaustion when it formulated what was then the equivalent of Art 8(2), but was also concerned with justifying the extension of protection for patented processes that enables the production of biological material to products other than those which are the direct products of the patented process. In this sense, Art 8(2) is to process patents what Art 9 is to product patents.

On the other hand, Kock's assertion that the reason for the inclusion of the requirement to 'perform its function' was to confirm that the patentee's rights are exhausted when the genetic information has (due to propagation) lost its functionality finds little support in either the negotiating history of the Directive, or the debates of the Committee of Experts. Doubtless, the drafters of the Directive were concerned with ensuring that the doctrine of exhaustion was applied no differently to biotechnological inventions, particularly those having the capacity to self-reproduce,[170] than it is to inanimate inventions. The fact that inventions of

---

168   Ibid., 166 (Amendment Nos 29 and 30).
169   Ibid., 167 (Amendment No. 31).
170   See Roberts, T. 1995. Draft Directive on legal protection of biotechnical inventions. *European Intellectual Property Review*, 17(5), D116–17. Commenting on the Amended Original Proposal, Roberts notes that Arts 9 to 11 have been contentious:

this kind, when used for the purposes for which they are intended, provide the means for further reproduction of the invention makes patents for such inventions especially vulnerable to infringement. However, it is far from clear that Art 9 is exclusively, or even primarily, concerned with exhaustion. This much is apparent from the negotiating history of the Directive.

When the provision on which Art 9 is based was being debated by the Committee of Experts, the primary concern was to establish the grounds on which *de facto* protection of products which incorporate patented genetic information is justified. As described above, the initial proposal prepared by the International Bureau did not consider that this *de facto* protection was warranted by the mere presence of the genetic information in the relevant product. Rather, this extension of protection was justified only where the genetic information is of essential importance for the industrial applicability of the product. In other words, conferring *de facto* protection to a product that is not the subject of a claim as such, but which contains genetic information that is the subject of a claim, is justified insofar as use of the product involves use of the protected genetic information. Until the basis for the extension of protection had been established, it was not possible to determine what rights a claim to genetic information provides the patentee with and, thus, when and how those rights might be exhausted. Each alternative version of Art 9 discussed by the Committee of Experts was also predicated on this assumption and, given that both of these views were reproduced by early drafts of the Directive, its drafters must be taken to have accepted this.

This is entirely in accordance with general principles. As Amiram Benyamini observes:

> [w]hile it is plain that the scope of protection is limited to the product claimed, it must also be stressed that the use or sale of a particular article is not exempted from infringement by the mere reason that it is completely different from the patented product, if it contains that product. A patented product is protected as such whatever purpose it serves ... Accordingly, it is an infringing act to sell or use cars containing a patented component, pharmaceuticals or substances containing a patented ingredient, products made of a patented raw material, and even buildings constructed with patented materials.

> The application of this rule depends on the existence of the patented product in the end-product. It means, in practice, that the patented product must still retain its distinct entity [sic] *and main properties* after incorporation into the

---

'Two well-recognised principles of patent law in this instance clash: the first, that a patent must give the right to control multiplication of what is claimed; the second, that the purchaser of an invention from the patentee normally expects the right to use it for its intended purpose and deal with it free of further constraints ... the Directive confirms that, in this case, the former principle prevails.'

greater article, and can be isolated or identified as such. Moreover, *the patented product must function in the incorporating article in the manner envisaged by the patent, so as to contribute in some way to the functioning of the article.* It is not sufficient for infringement that the incorporating article contains some remains, 'finger prints' or 'impurities' of the patented product, *which have no effect on its functioning.* Where [these] two conditions ... are not fulfilled, it would be fictitious to say that the 'product which is the subject-matter of the patent' is used or sold as part of the greater article, and that using or dealing with that article constitutes use of the invention.[171]

Benyamini, now a District Court Judge in Tel-Aviv, confirms that '[w]here the patented product exists and fulfils its patented function in the greater article', liability for infringement of the product is predicated on the fact that 'any use or dealing with that article constitutes inevitably use of the invention.'[172] Thus, rather than being seen as a discriminatory gesture towards biotechnological inventions, Art 9 reaffirms the principles of infringement relating to subject matter that is 'intended to be, by its nature, a component or ingredient of another product'.[173]

Other commentators who discussed this issue around the time the Directive was being negotiated expressed a similar view. Writing shortly before the Original Proposal was introduced, Bernhard Roth, then Vice-President of Ciba-Geigy Ltd, observed:

It is generally accepted that [for example] a new man-made gene is patentable and that such a gene incorporated into a plant is part of that plant. Even if a

---

171   Benyamini, above n 149, 97–8 (emphasis supplied).

172   Ibid., 98.

173   Benyamini acknowledges that the application of these rules may cause difficulties in the field of biotechnology, and at one point suggests that 'when a patented gene is inserted into the plant genome, it is arguable that any future generation or embodiment of that plant, a cell, tissue, part of a plant (a flower or fruit), a seed, and even a food product which contains the patented gene might be seen as infringements of a claim to the gene,' ibid., 98–9. However, this view is based on the assumption that 'the patented gene exists and contributes to the end-product', a presumption which emerges more clearly when he discusses Art 13 of the Original Proposal. There, he asks whether or not it will be an infringement 'to sell a fruit jam containing the patented gene of a new fruit? If the new fruit is essential to the quality of the jam the answer may well be positive,' ibid., 100 (n 13). Likewise, Straus has argued that products derived from a plant containing patented genetic information, for instance, polenta from Bt-corn, should only be regarded as an infringement where the products owe their characteristics to the invention – for example, where a gene used for the transformation of maize plants is 'responsible for a specific nutritional value, or taste, or consistency of maize grain, from which polenta/semolina were produced by further processing and where polenta would retain the specific nutritional value, taste or consistency'. Of course, in this situation claims to these derivative products themselves could be obtained, a course that, despite the suggestions of some commentators, was not available to Monsanto. See Straus, above n 149, 654.

second, third, fourth or tenth generation of this plant exists, if it still contains the new gene and the plant is being used, *then the gene is also being used*. If a patented gene is incorporated into the new plant *the benefit of the invention will have been taken* and the new plant or its propagation material should therefore be considered an infringement of the gene patent.[174]

Similarly, in its response to the questionnaire circulated by WIPO prior to the third meeting of the Committee of Experts in 1987, the AIPPI asserted that the protection given to patented genetic information 'should extend to plants or animals incorporating the patented biological material where the material contributes significantly to the properties or characteristics of the plant. In such circumstances commercial activities in relation to the plant or animal *would involve use of the patented biotechnological material*'.[175] More recently, Joseph Straus, a long-standing supporter of strong patent protection for biotechnological inventions, who was involved in the preparation of the Original Proposal and later participated as an expert in the hearings held by the European Parliament Committee for Legal Affairs and Citizen's Rights in 1996, has argued that 'the extent of protection under Art 9 does not depend on whether the respective patented product [i.e. the greater article] was used, but on the fact that whether *by use of that product also the genetic information incorporated therein was used*. In other words, the respective patented product was used because of the specific function of the respective genetic information disclosed in the patent application.'[176]

---

174   Roth, B. 1987. Current problems in the protection of inventions in the field of plant biotechnology – A position paper. *International Review of Industrial Property and Copyright Law*, 18(1), 41–55, 53. Both the title of his paper and the language chosen by Roth leave no room for doubt that his is a normative position. This is confirmed by Andrew Christie who, writing shortly after Roth, noted that '[r]esolution of this issue [i.e. whether incorporation of the genetic information into another plant constitutes an infringement of patent] requires a policy decision'. Christie, above n 100, 408. Christie approves of the position adopted by the Original Proposal in this regard as 'a suitable approach'.

175   International Association for the Protection of Intellectual Property. 1987. WIPO Questionnaire Concerning the Legal Protection of Biotechnological Inventions, Document BIOT/Q2: Response on Behalf of AIPPI. *Annuaire* 1987, 1, 81–4, 82. [Online]. Available at: https://www.aippi.org/download/yearbooks/Annuaire%201987_I.pdf [accessed 21 October 2013] (emphasis supplied).

176   Straus, above n 149, 647–8 (emphasis in original). In the same paper, Straus suggests that a hypothetical claim to genetic information which imparts herbicide or drought tolerance upon maize plants in which it is expressed would not be infringed under Art 9 by processing of the plants into polenta because the latter product 'does not contain any genetic information which would make polenta herbicide resistant or drought tolerant'. This would be the case even where the polenta contains residues of the patented genetic information. Nor, in his opinion, would claims to the polenta be allowable, 'since neither the polenta nor the soymeal at hand owe any of their characteristics to the invention'. The situation would

Therefore, Art 9 might be better understood as an instantiation, in the context of biotechnological inventions, of the scope of protection for product claims where the subject matter of the claim must, by its nature, become a part of another product in order for it be exploited.[177] In the same way that exhibiting a motor vehicle fitted with patented tyres,[178] or transporting bottles of beer sealed with a patented capsule,[179] or keeping patented pumps on a ship,[180] gives rise to liability for infringement because the invention is subserving the purpose for which it was patented, so too the use of a plant variety that incorporates patented genetic information – for example, for the purpose of producing another variety – will attract liability for infringement whenever the genetic information performs its function (typically, the function(s) disclosed in the specification) or remains capable of doing so.

The same logic underlies the decision of the Supreme Court of Canada in *Monsanto Canada, Inc. v. Schmeiser*.[181] There, a majority held that the Saskatchewan canola farmer, Percy Schmeiser, infringed a claim of Monsanto's patent for a gene encoding glyphosate herbicide resistance on plant cells in which it is incorporated, notwithstanding the fact that Monsanto was unable to establish that Schmeiser had in fact treated his crops with Roundup or any other glyphosate herbicide: '[c]ase law shows that infringement is established where a defendant's commercial or business activity involving a thing of which a patented part is a

---

be different, however, if the genetic information conferred 'a specific nutritional value, or taste, or consistency of the maize grain, from which polenta/semolina were produced by further processing and where polenta would retain the specific nutritional value, taste or consistency'. Despite this, Straus maintains that Arts 8 and 9 of the Directive 'seem adequate and well adjusted to the needs of those who innovate in the plant area' (653–6).

177   Some support for this view can be found in Recital 8, which provides, among other things, that 'legal protection of biotechnological inventions does not necessitate the creation of a separate body of law in place of the rules of national patent law'. See also Llewelyn and Adcock, above n 68, 378: 'it is common patent practice for the claims to extend to any material within which the patented invention is placed or utilised. Articles 8 and 9 of the Directive reiterate both of these principles and attach them firmly to biological material' (emphasis supplied); Bostyn, S. 2004. *Patenting DNA Sequences (Polynucleotides) and Scope of Protection in the European Union: An Evaluation.* Brussels: European Commission, 108–9. [Online] Available at: http://ec.europa.eu/internal_market/indprop/docs/invent/patentingdna_en.pdf [accessed 14 March 2014]: '[Arts 8 and 9] are new in the sense that they were previously not found in most national patent acts. They do not create new rules, however, since similar provisions were developed in the past by interpreting the existing traditional patent law provisions. The advantage, at least at first sight, of these new provisions, is that they codify practices and policies that have been developed over time in court decisions.'

178   *Dunlop Pneumatic Tyre & Co. Ltd v. British & Colonial Motor Co. Ltd* (1901) 18 R.P.C. 313.

179   *Betts v. Neilson* [1868] L.R. 3 Ch. App. 429.

180   *Adair v. Young* (1879) 7 Ch. 13.

181   [2004] 1 S.C.R. 902.

component *necessarily involves use of the patented part.'*[182] Although Schmeiser did not actively exploit the invention (by treating his crops with Roundup), the majority nevertheless found that he had benefited from the so-called 'stand-by utility' of the invention – that is, the liberty to use the invention should the occasion require. This is in keeping with older English authorities that establish that possession of a patented invention in the course of trade creates a rebuttable presumption that the defendant intends to use it, and thereby becomes liable for infringement – or subject to a *quia timet* injunction – even in the absence of evidence of actual use.[183]

However, it appears that liability for infringement under the stand-by utility doctrine is limited to situations where the invention is adapted for ready use.[184]

---

182   [2004] 1 S.C.R. 902, 904. The majority analogised the patented genes, and cells containing them, to patented Lego blocks: 'the patented genes and cells are not merely a "part" of the plant; rather, the patented genes are present throughout the genetically modified plant and the patented cells compose its entire physical structure. In that sense, the cells are somewhat analogous to Lego blocks: if an infringing use were alleged in building a structure with patented Lego blocks, it would be no bar to a finding of infringement that only the blocks were patented and not the entire structure. If anything, the fact that the Lego structure could not exist independently of the patented blocks would strengthen the claim, underlining the significance of the patented invention to the whole product, object, or process' (921–2). As discussed earlier, this resembles the justification given by the Commission for the original version of Art 9 (Art 13 of the Original Proposal). However, Pottage and Sherman criticise this reason for blurring the 'essential distinction between tangible and intangible, or between kind and embodiment, and in so doing [unravelling] the basic scheme of patent law … What counts is the reproduction of the idea, not the physical material in which its embodied'. See Pottage, A. and Sherman, B. Biological and industrial kinds: Plants as intellectual property. Unpublished paper on file with author, 37.

183   The leading decision is *British United Shoe Machinery Co. v. Simon Collier* (1909) 26 R.P.C. 21 (Parker J.); (1909) 26 R.P.C. 534 (Court of Appeal); (1910) 26 R.P.C. 567 (House of Lords). There, the defendants were alleged to have infringed a patent for a mechanism for producing a particular type of sole on shoes ('Scotch soles'). The defendants had purchased a machine for trimming the soles of shoes that was fitted with the device, but the evidence established that they were not in the habit of producing Scotch soles and therefore had disengaged the patented mechanism. At first instance, Parker J. found for the patentee. This decision was reversed by the Court of Appeal, and their decision was upheld by the House of Lords. See also *Badische Anilin und Soda Fabrik v. The Basle Chemical Works, Bindschedler* [1898] A.C. 200, where Lord Herschell reserved the question of whether mere possession of a patented article amounts to use of the invention in circumstances where the defendant derives no advantage from it; and, *British Motor Syndicate Ltd v. Taylor & Son* [1901] 1 Ch. 122.

184   See *British United Shoe Machinery Co. v. Simon Collier* (1909) 26 R.P.C. 534, 541 (Fletcher-Moulton LJ). It is notable that the two examples of stand-by use referred to in this decision are articles fit for immediate use, namely a fire extinguisher and a spare steam engine. Philip Grubb concurs with the view that stand-by use is applicable to situations where the patented article is in a position ready for use. See Grubb, P. 2005. *Patents for Chemicals,*

Indeed, on several occasions the majority in *Schmeiser* treat 'stand-by' as synonymous with 'insurance', signifying that the invention is ready-to-hand. This is the critical difference between the decision of the Supreme Court of Canada and the CJEU: in the former case, the invention's stand-by utility was actualised by spraying the crop with herbicide; in the latter, the genetic information could only be utilised after first being extracted from the soy meal and purified, and then re-introduced into a suitable transformation vector and transferred to a plant cell. This is hardly a straightforward or routine exercise. Nor was any evidence produced by Monsanto to show that the defendants were in the habit of performing this exercise, or intended to do so. However, it does explain why Monsanto pursued so earnestly the argument that it is sufficient, for the purposes of establishing liability under Art 9, to show that the protected genetic information has performed its function or is capable of doing so once again.

On this interpretation of Art 9 no violence is inflicted on the principle of absolute protection. Where the patented product exists and fulfils its patented function in the greater article, any use or dealing with the article inevitably involves use of the invention. This will ordinarily be the case. As Lord Justice Stephenson noted in *R v. Patents Appeal Tribunal; ex parte Beecham Group Ltd*, '[i]n the ordinary understanding of words he who uses a thing made up of different ingredients uses them all, whether he knows what each is and even if he intends to use the thing but not a particular ingredient in its composition because he does not know of its existence.'[185] However, where the defendants are not actively exploiting the advantage obtained by the invention, there is little justification for making them liable for infringement.

Although Kock's interpretation of Art 9 provides one answer to the interpretative conundrum identified by the CJEU, he makes no attempt to reconcile Art 9 with Art 8(1).[186] Indeed, it is difficult to identify in what circumstances Art 9 would operate where Art 8(1) does not. According to the definition in Art 2(1)(a), genetic

---

Pharmaceuticals and Biotechnology: Fundamentals of Global Law, Practice and Strategy, fourth edition. Oxford: Oxford University Press, 156. See also *Beecham Group Ltd v. Bristol Laboratories Ltd* [1978] R.P.C. 153, where the defendants were found liable for infringing a patent covering the antibiotic ampicillin by importing into the United Kingdom the pro-drug hetacillin. Hetacillin is formed by the addition of acetone to ampicillin, to which it reverts following ingestion. Although the case was treated as an instance of infringement of the 'pith and marrow' of the patented invention (hetacillin being ampicillin 'temporarily masked'), it is also explicable on the grounds that the defendants benefited from the stand-by utility of ampicillin, which could readily be obtained by means of a non-technical process.

185   [1973] 1 Q.B. 318, 334.

186   Llewelyn and Adcock have recognised this problem, although the solution they offer is far from satisfactory. In short, they suggest that infringement under Art 8(1) arises irrespective of whether the protected genetic information performs its function (passive infringement), whilst infringement under Art 9 is established only when the protected genetic information performs its function (active infringement). However, where the genetic information is not performing its function in subsequent generations of biological material,

information is clearly a kind of 'biological material', being material *containing* genetic information and capable of reproducing itself or being reproduced in a biological system. Whilst it may seem tautologous to suggest that genetic information *contains* genetic information, as the Explanatory Memorandum to the Second Proposal points out, 'the concept of genetic information automatically makes reference to a material substratum on which it is based, namely deoxyribonucleic acid.'[187] In other words, DNA contains genetic information.[188] Moreover, the expression 'the protection conferred by a patent on a biological material possessing specific characteristics as a result of the invention' should be read in the extended sense; that is, 'the protection conferred by a patent' includes the product given *de facto* protection by Art 9.

Absolute or Purpose-bound Protection?

There appears to be little justification for inferring from the provisions of the Directive or its negotiating history an intention to interpret claims to DNA sequences as being limited to the functions described in the specification. Advocate General Mengozzi arrived at this interpretation in the course of dispensing with the Italian Government's argument that 'classic patent protection' ceases once genetic information is incorporated in a greater product, whereupon the only protection available is that provided by Art 9.[189] The rejection of this argument compelled Professor Mengozzi to reconcile the apparent conflict between the principle of absolute protection and Art 9. He attempted to do so by emphasising certain provisions of the Directive that, in his view, 'permits – and, in fact, requires – an interpretation to the effect that … the protection conferred on DNA sequences is a purpose-bound protection'.[190] Professor Mengozzi acknowledged that 'the Directive does not expressly indicate that the protection conferred on DNA sequences must be of that order,' and that the provisions of the Directive on which he relies for this interpretation 'concern the scope of patentability rather than the scope of the protection for the patented product'.[191] Nevertheless, the 'great importance attached by [the Directive] to the function performed by a DNA sequence', which 'is naturally intended to permit a distinction to be drawn between "discovery" and "invention"', compelled the Advocate General to arrive at this conclusion:

---

it is highly unlikely that these materials will possess the 'specific characteristics' that are imparted to the material by the invention: Llewelyn and Adcock, above n 68, 378–80.

187   COM(95) 661 final – 95/0350(COD), above n 119, 19.

188   'The order in which the four bases A T G C occur constitutes the genes' encoded information.' This interpretation is also supported by the legislative history relating to the definition of 'biological material' and 'self-replicating material'. European Parliament, COM(95) 661 final, ibid., 19.

189   Case C-428/08, above n 34, I-6776–7, [25].

190   Ibid., I-6777, [29].

191   Ibid., I-6778, [30].

The isolation of a DNA sequence without any indication of a function constitutes a mere discovery and as such is not patentable. Conversely, the sequence is transformed into an invention, which can then enjoy patent protection, through the indication of a function that it performs. However, to maintain that a DNA sequence enjoys 'traditional' patent protection [i.e. absolute protection] ... would mean that patents would be recognised as covering functions as yet unknown at the time of the patent application. In other words, lodging an application for a patent for a *single function* of a DNA sequence is all it would take to obtain protection for *all the other possible functions* of the same sequence. In my view, such an interpretation would ultimately, in practice, make a mere discovery patentable, in breach of the basic principles on patents.[192]

The Grand Chamber did not express an opinion on this interpretation. However, Professor Mengozzi might have dispensed with this argument on the general principles of patent law described above. Historically, the language of absolute protection has been invoked in the context of debates about whether the scope of protection for chemical products should embrace all possible methods of manufacture and, more recently, in the context of the debates about whether claims to DNA should encompass only those uses identified by the patentee in the specification, or whether they should embrace what might be termed 'prophetic' uses of the invention about which the specification is silent. Obviously, the litigation before the CJEU is far removed from either of these scenarios.

Furthermore, Art 5(3) of the Directive, which requires the industrial applicability of a sequence or partial sequence of a gene to be disclosed in the patent specification, merely restates the general principle that the specification shall disclose the industrial applicability of the invention (where this is not obvious from the specification)[193] and was only inserted in the Amended Second Proposal introduced late in 1995 in response to the controversy associated with the now infamous applications filed by the National Institutes of Health in the US and the British Medical Research Council in the early 1990s, covering thousands of

---

192   Ibid., I-6778, [31] (emphasis in original).

193   This is confirmed by Amendment 16 recommended by Parliament, which introduced 5 new Recitals – 16(a) to 16(e) – that were included by the Commission in the Amended Second Proposal. Recital 16(b) provided that 'a mere sequence of DNA segments does not contain a technical teaching and is therefore not a patentable invention.' In the Commission's Amended Second Proposal 'technical teaching' was treated as synonymous with 'biological function'. Recital 16(a) provided that 'the granting of a patent for inventions which concern such sequences or partial sequences require the same criteria to be applied as in all other areas of technology,' and Recital 16(e) confirmed that 'the requirements for disclosure of the industrial application of the sequences or partial sequences do not differ from those in other areas of technology' and 'at least an industrial application must be actually disclosed in the patent application'.

expressed sequence tags of questionable utility.[194] Indeed, by requiring disclosure of the biological function of a gene, rather than its *technical* function, Art 5(3) may be more generous to applicants than the general patent law. Although there are cogent *a priori* reasons to question the legitimacy of *per se* claims to genetic information, these reasons could equally be used against chemical products generally, and yet here absolute protection remains the norm. To be sure, the functional promiscuity of genetic information might suggest that these criticisms have more relevance in the context of biotechnological inventions, where functional diversity may be the rule, than elsewhere. Nonetheless, if the drafters of the Directive had intended to subject product claims to genetic information to purpose-bound protection, one would ordinarily have expected them to express this intention with greater clarity than it has purportedly done – for example, by clearly stating that the scope of protection for genetic information is limited to the industrial application claimed, rather than described.

Conclusion

Given the tumultuous history of the Directive, it was almost inevitable that the first decision to interpret its substantive provisions would raise as many questions as it answered. Indeed, the commentary that has accompanied the CJEU's decision rehearses the, at times bitter, divisions that marked the negotiation of the Directive. From both perspectives, the Directive embodies a compromised, rather than ideal, position. It may well be, as Jan Krauss and Toshiko Takenaka have argued, that the CJEU's decision 'is a response to the political pressure coming from European politics'.[195] Support for this interpretation might be found in the European Parliament's 2005 resolution on patents for biotechnological inventions, which called on the European Patent Office and Member States 'to grant patents on human DNA only in connection with a concrete application and for the scope of the patent to be limited to this concrete application so that other users can use and patent the same DNA sequence for other applications (purpose-bound protection)'.[196] Already, several European countries have, of their own volition, moved in this direction, including France, Germany, Switzerland, and Italy.

The largely critical reaction to the CJEU's judgment might also be born of disbelief that patent law, at least in this situation, provides more limited protection

---

194 See Parliament's Amendments 100 and 49. See also Crespi, R. 1995. The European Biotechnology Directive is dead. *Trends in Biotechnology*, 13(5), 162–4, 163.

195 Krauss and Takenaka, above n 68, 201.

196 European Parliament. 2005. Patents on Biotechnological Inventions: European Parliament Resolution on Patents for Biotechnological Inventions. *Official Journal of the European Communities* C 2/2E, 440, 441, [5]. As Parliament notes, however, this may only be achieved by means of a recommendation to Member States or by amending Art 5. In other words, this interpretation is not open according to the Directive as written.

than plant variety rights – in particular, the 1991 iteration of the UPOV Convention, with its extension of the scope of protection to harvested material and products derived from harvested material – which has hitherto been regarded as offering less extensive protection than patent law. Likewise, the importation of products produced abroad using a patented process has been a perennial concern of patent law. It was these concerns which led English courts in the late nineteenth and early twentieth centuries to extend the scope of protection for process patents to products produced overseas using the patented process and imported into the jurisdiction.[197] From this perspective, the CJEU's decision is a retrograde step, rather than the step forward many expected once the Directive was introduced.

There is little justification, however, for some of the more hysterical responses to the CJEU's decision. It is not the case, as some have argued, that the decision puts at risk 'the enforceability of patent claims directed to isolated nucleotides used as reagents, such as reagents in diagnostic methods ... [because] these nucleotides do not perform their function in a reagent vial or kit'.[198] To so read the judgement is to wilfully ignore the court's interpretation of 'function' as that which is disclosed in the specification, as well as the basic principle that the specification must disclose 'in definite technical terms the purpose of the invention and how it can be used in industrial practice to solve a given technical problem'.[199] This interpretation is also incongruous with the terms of Art 9 itself: when used as reagents, it is not necessary for isolated nucleotides to be incorporated within another product in order to perform their function *as reagents.*

Nor is it the case that the Court 'applied a "function-limited" or "purpose-bound" protection for DNA sequence patents',[200] or that its judgement can be read as providing 'an exception to the general principle that product claims confer absolute protection'.[201] At the present time, this interpretation of the scope of claims to DNA sequences partakes more of a normative aspiration, predicated on past experience in relation to chemical compounds, than of one grounded in precedent. As much is conceded by Krauss and Takenaka, who, despite arguing that the CJEU's interpretation of Art 9 is 'clearly in conflict with the current case law of many EPC members states such as Germany', acknowledge that 'there is no case law of the [German Federal Court of Justice] specifically related to DNA

---

197   However, where claims to products are concerned, English courts have always been loath to grant too wide a berth to the extra-territorial application of English patent law. See *Wilderman v. F.W. Berk & Co. Ltd* [1925] 1 Ch. 116; *Beecham Group Ltd v. Bristol Laboratories Ltd* [1978] R.P.C. 153; Lynfield, H. 1965–1966. Infringement in Great Britain by importation of transformed products. *Patent, Trademark and Copyright Journal of Research and Education,* 9, 577–86.

198   Mohan-Ram *et al.* above n 156, 551. See also Kock 2010b, above n 132, 755.

199   T0898/05 *Hematopoietic receptor/Zymogenetics* [2007] E.P.O.R. 2, [6].

200   Krauss and Takenaka, above n 68, 193.

201   See Kock 2010b, above n 132, 755; Morgan, G. and Haile, L. 2010. A shadow falls over gene patents in the United States and Europe. *Nature Biotechnology,* 28(11), 1172–3; Morgan, G. 2011. Devil in the genetics. *European Lawyer,* 31–2.

sequences.'[202] Indeed, the CJEU's decision confirms a trend among European courts, particularly those in the United Kingdom, which suggests that the scope of protection for biotechnological inventions may not be as broad as first thought.[203] As Christopher Holman has observed, this is no less true in the US, at least with respect to litigation involving human gene patents.[204]

While critics of the decision have, understandably, emphasised the potential consequences of the CJEU's decision for the biotechnology industry, exploration of the consequences of Monsanto's claims for other participants in the food industry, including consumers, has been left to others. The impossibility of specifying 'for how long, or up to what stage of the food and derived product chain, traces of the original DNA of the [GM] plant are still identifiable' was a crucial factor in the Advocate General's rejection of Monsanto's claim. If accepted, this 'would ultimately lead the holder of a biotechnological patent to be granted too wide a range of protection'.[205] As the legal representatives for Cargill in the English litigation argued, taken to its logical conclusion Monsanto's argument would mean that 'any product containing an intact DNA molecule derived from any GMO would [amount to] an infringement of any patent right directed at those gene sequences used to construct that [GMO].'[206]

---

202   Krauss and Takenaka, above n 68, 193, note 15.

203   In the United Kingdom see *Biogen Inc. v. Medeva Plc.* [1997] R.P.C. 1; *Kirin-Amgen Inc. v. Hoechst Marion Roussel Ltd* [2005] R.P.C. 169. See also Price, R. and Morgan, G. 2009. Enforcing DNA product claims in Europe: A case study using Monsanto's Roundup Ready litigation. *European Intellectual Property Review*, 31(7), 377–84.

204   See Holman, C. 2008–2009. Learning from litigation: What can lawsuits teach us about the role of human gene patents in research and innovation. *Kansas Journal of Law and Public Policy*, 18(2), 215–72, at 223–4: 'When considering the impact of human gene patents, commentators often seem to assume that the existence of a human gene patent effectively blocks all practical uses of the patented gene, but this is far from the truth. Most of the patents asserted in the context of therapeutic proteins do not claim the gene per se but only include claims limited to some specified compositions of matter comprising the human gene or a method using the human gene to produce a protein encoded by the gene. Only three of the twenty-one patents asserted in this context included an isolated polynucleotide claim, the prototypical gene patent claim most people think of as a gene patent. In two cases – *Amgen v. Chugai Pharmaceutical Co.* (927 F. 2d 1200 (1991)) and *Genentech v. Insmed* (436 F. Supp. 2d. 1080 (2006)) – a court held that the isolated polynucleotide claim had been infringed. In the only other case involving an isolated polynucleotide claim, *Genentech v. Wellcome Foundation* (29 F. 3d. 1555 (1994)), the court held that the isolated nucleotide claim was not infringed – an outcome which serves to illustrate how even an apparently broad claim purporting to encompass any recombinant use of the gene can nonetheless be susceptible to technical circumvention.' See also Holman, C. 2007–2008. The impact of human gene patents on innovation and access: A survey of human gene patent litigation. *University of Missouri-Kansas City Law Review*, 76(2), 295–361, 302–3.

205   Case C-428/08, above n 34, I-6779, [34].

206   Cohen and Morgan, above n 143, 291.

The implications of this reasoning are as sensational as they are startling. Before the CJEU, the Argentine government argued – 'following a line of reasoning that is', in Professor Mengozzi's view, 'paradoxical only in part' – that 'if traces of the sequence were to be detected in the stomachs of cattle because the animals had been fed with products derived from the genetically modified plant, even the importation of those cattle could be regarded as an infringement of the patent-holder's rights.'[207] In a similar vein, Christopher Heath, a member of the EPO's Boards of Appeal, has stated that:

> If Monsanto's position were correct, similar cases could arise for the importation of cotton, the RR variety of which Monsanto has started selling in India without corresponding patent protection. Any trace of the protected DNA in, say, imported jeans or t-shirts would allow the patentee to raid the premises of any manufacturer or shop commercially manufacturing or selling these goods – an interesting, but somewhat worrying scenario.[208]

The CJEU litigation might best be understood, then, as another instantiation of the adage made popular by Oliver Wendell Holmes, that 'hard cases make bad law.'[209] Nearly 150 years ago, the English Lord Chancellor declined to engage in a hypothetical put to him by the defendant's counsel in a case involving infringement of a patented capsule affixed to bottles of beer. The bottles were carried as cargo on a ship that temporarily entered the United Kingdom, pending their export to another country. Counsel put it to Lord Chelmsford that if the defendant was liable, then so too was a woman who purchased a dress in Paris which had been coloured with a dyestuff patented in England and worn into the country. Such 'an extreme case' was, according to his Lordship, 'merely a matter for ingenious speculation, and not even likely to arise in such a manner as to call for a legal decision'.[210]

Perhaps this was a signpost to the outcome of this litigation. Indeed, Kock has chided Monsanto for 'over-aggressively asserting their patent rights' and, in the process, imperilling 'the investment strategies of entire industries',[211] while another of the most prominent critics of the CJEU's decision, Gareth Morgan, has conceded that there is a 'feeling across the EU ... that Monsanto should have fought this battle in South America'.[212] Correa goes one step further, describing

---

207   Ibid.

208   Heath, above n 31, 956. See also Correa, above n 14.

209   *Northern Securities Co. v. United States*, 193 U.S. 197, 400 (1904).

210   [1868] L.R. 3 Ch. 429, 438. Of course, so far as intra-European trade is concerned, the doctrine of exhaustion forecloses the possibility of infringement in these circumstances.

211   Kock. 2010b, above n 132, 756.

212   See Morgan, above n 201, 32. See also Case C-428/08, above n 34, I-6779, [35]: 'There is no doubt that the lack of protection for Monsanto's invention in Argentina seems unfair. By the same, token, however, and leaving aside the reasons for that lack of protection, it seems to me that Monsanto's plan of action is to try and use one legal order

the initiation of these proceedings by Monsanto as an exemplary form of 'strategic litigation' in which the plaintiff 'has little chances of success or no real interest in obtaining a final judgment' but instead commences legal action in order to 'put pressure [on the defendant] and create risks and uncertainty, thus achieving a rapid subordination of targeted parties to commercial conditions with no legal basis'.[213] As much was admitted by Monsanto itself. When negotiations with the Argentine government broke down, Monsanto affirmed its commitment to seek 'a local solution and a local agreement' with the Argentinian government, before reaffirming, after the CJEU's decision, its belief that 'the correct place for a resolution of these matters is in Argentina'[214] – a position that is difficult to reconcile with its actions. The purpose of the suits, a company official said, was merely to support the company's claim that it has a legal right to collect royalties, if not in Argentina then at the point of delivery.[215] Indeed, in 2008, whilst the CJEU was still in the process of determining the case, the firm announced its intention to invest US$125 million in Argentina over the next four years,[216] and signalled its intention to introduce a new GM soy cultivar, known as 'Intacta RR2 Pro', in Argentina in 2014, two years after its introduction in Brazil and Paraguay. Despite the biotechnology industry's discontent with the CJEU's decision, for Monsanto and Argentina this love story may yet have a happy ending.

Table 3.1   Genealogy of Art 9

WIPO Committee of Experts on Biotechnological Inventions and Industrial Property	
Suggested Solution 13	*'It appears desirable that patent protection for a product consisting of, or containing, particular genetic information as an essential characteristic of the invention should extend to any products containing the patented product or obtained from such product in which the said genetic information is of essential importance for the industrial applicability or utility of the said product.'*
Revised SS 13 (SS 9)	*'Patent protection for a product that consists of, or contains, particular genetic information as an essential feature of the invention shall extend to any matter containing the patented product or obtained from the patented product provided that the said genetic information is contained and expressed in the said matter.'*

(that of the European Union) to remedy problems encountered in another legal order (that of Argentina). That seems to me, however, to be unacceptable. The fact that Monsanto cannot obtain adequate remuneration for its patents in Argentina cannot be remedied by according Monsanto extended protection in the European Union.'

213   Correa, above n 13.

214   Monsanto, above n 65.

215   Turner, above n 20.

216   'Argentina: Monsanto introduces new form of genetically modified soybean seed,' *La Nacion*, 28 August 2008.

Table 3.1   *Continued*

Further revised SS 13	*'Patent protection for a product that consists of, or contains, particular genetic information as an essential feature of the invention as defined in the claim or claims shall extend to any use of the patented product as contained in any matter provided that the said genetic information is essential for its use.'*

Biotechnology Directive, Original Proposal

Art 13	*'The protection for a product consisting of or containing particular genetic information as an essential characteristic of the invention shall extend to any products in which said genetic information has been incorporated and is of essential importance for its industrial applicability or utility.'*

Biotechnology Directive, Amended Original Proposal

Art 12	*'The protection conferred by a patent on a product containing or consisting of genetic information shall extend to all material in which the product is incorporated and in which the genetic information is contained and expressed.'*

Biotechnology Directive, Second Proposal

Art 11	*'The protection conferred by a patent on a product containing or consisting of genetic information shall extend to all material, save as provided for in Art 3(1), in which the product is incorporated and in which the genetic information is contained and expressed.'*

Biotechnology Directive, Amended Second Proposal

Art 9	*'The protection conferred by a patent on a product containing or consisting of genetic information shall extend to all material, save as provided for in Art 5(1), in which the product is incorporated and in which the genetic information is contained and expressed.'**

Biotechnology Directive

Art 9	*'The protection conferred by a patent on a product containing or consisting of genetic information shall extend to all materials save as provided in Art 5(1), in which the genetic information is contained and performs its function.'*

Note: * Although it involves no substantive change from its previous incarnation in the Second Proposal, Art 9 is said to have been 'inspired by amendment 58'. Amendment 58 included a reference to both Art 3(1) and Art 2a(1). The latter provided that plant and animal varieties, and essentially biological procedures for the breeding of plants and animals, shall not be patentable. This was a new provision introduced by Parliament by way of amendment 47. Although this amendment was incorporated by the Commission in Art 4, the Commission removed Parliament's reference to Art 2a(1) on the grounds that it was 'technically and legally incomprehensible' in view of Arts 11 and 12 and would in practice limit the scope of protection conferred by a patent 'in a such a way as to go against current practice under patent law'. See Commission of European Communities, 'Amended Proposal for a European Parliament and Council Directive on the Legal Protection of Biotechnological Inventions', COM(97) 446 final, [1997] *Official Journal of the European Communities* C 311/12, 5.

Chapter 4

Competition in the Agricultural Seeds Sector: Patents and Competition at a Cross-roads?

Charles Lawson

Introduction

In a developed capitalist economy markets and competition are supposed to allocate resources to best satisfy the needs and requirements of consumers. Where a firm gains market power there is the potential to abuse that power and undermine the positive operation of markets and competition to the detriment of consumers.[1] The development and adoption of GMOs in agriculture is likely to be a significant node of problematic competition because of the tensions between patent law and antitrust law. Patent law was expanded in *Diamond v. Chakrabarty* to include GMOs in 1980[2] and then in 2001 in *J.E.M. Ag Supply, Inc. v. Pioneer Hi-Bred International, Inc.* (confirming the decision in *Ex parte Hibberd*)[3] to include traditionally bred non-genetically modified plants.[4] This raises the possibility that many of the practices using plants that are now conducted in the gaze of patents were earlier only regulated by competition laws. The effect of these decisions has been to extend the scope of utility patents and eliminate some of the previously pro-competition limitations – like farmer saved seeds, research exemptions and exhaustion provisions – that were available without patents. Add to this the increasing involvement of the for-profit private sector taking over from the public sector in developing new plants,[5] the consolidation of the seed

---

1   In the agricultural context see, for example, Domina, D. and Taylor, R. 2010. The debilitating effects of concentration markets affecting agriculture. *Drake Journal of Agricultural Law*, 15(1), 61−108, 62−3.

2   *Diamond v. Chakrabarty*, 447 U.S. 303, 305 (1980).

3   *Ex parte Hibberd*, 227 U.S.P.Q. 443 (Board of Patent Appeals, 1985).

4   *J.E.M. Ag Supply, Inc. v. Pioneer Hi-Bred International, Inc.*, 534 U.S. 124, 148 (2001).

5   See, for example, Fernandez-Cornejo, J. 2004. *The Seed Industry in U.S. Agriculture: An Exploration of Data and Information on Crop Seed Markets, Regulation, Industry Structure, and Research and Development*, Agriculture Information Bulletin No. AIB-786. Washington: USDA, 42−6 (showing a 1,300 per cent real increase in private sector R&D plant breeding expenditures between 1960 and 1996 while, during the same period, public sector R&D plant breeding 'changed very little in real terms'). See also Heisey, P., King, J. and Day Rubenstein, K. 2005. Patterns of public-sector and private-sector patenting in agricultural biotechnology. *Ag Bio Forum*, 8(2−3), 73−82.

industry[6] and the potential for patent holders aggressively to assert their interests, and challenges to competition by firms developing and marketing GMOs can be expected. As a consequence the tension between patent laws and competition laws is likely to play out in the various markets involving GMOs.

The sub-markets for GMOs can conveniently be considered to be an innovation market for the discovery and development of novel and useful genetic traits, an innovation market for the discovery and development of novel and useful germplasm (plant breeding), a genetic traits market for the traits that can be placed into useful plant germplasm, a germplasm market for the germplasm that can have the traits inserted and a traited seeds market for the GMOs that can be planted and grown. Market power in each of these sub-markets and across these sub-markets enables firms exercising that power to foreclose competition by slowing innovation, raising prices, affecting quality, affecting choice and dulling the benefits of competition for producers (such as farmers) and the ultimate consumers of agricultural outputs. Of particular concern is the potential for patents to restrict consumer practices.

Resolving these patent and competition issues is important because developed economy agriculture depends on access to new technologies to improve agricultural production and yields. Perhaps significantly, market prices for agricultural outputs are generally set according to international benchmarks so that the likely effects of limiting competition will be a re-allocation of value within the market (principally through less profit for farmers)[7] and fewer choices for the ultimate consumers of agricultural production. The pricing of agricultural GMOs and the vast differential between GMOs and conventional seeds signals that patents and competition are an issue for GMOs. The US General Accounting Office, for example, concluded:

> Monsanto's US patents for Roundup Ready soybean seeds have given it and the companies to whom it has licensed the technology greater control over seed prices and has enabled them to restrict the availability and use of seeds … These factors do not have the same impact on Bt corn seed prices. Bt corn is genetically modified hybrid corn, and hybrid corn cannot be easily reproduced from seed. Thus, farmers and others cannot readily reproduce Bt corn seed for use on their farms.[8]

---

6   Hayenga, M. 1998. Structural change in the biotech seed and chemical industrial complex. *Ag Bio Forum*, 1(2), 43–55.

7   See, for example, Fernandez-Cornejo, above n 5, 9 (showing an increase from 1.6 per cent of total farm expenditures in 1960 to 3.7 per cent in 1997 or an increasing expenditure on seeds from US$1.95 billion in 1960 to US$ 4.81 billion in 1997, in 1989 dollars).

8   US General Accounting Office. 2000. *Report to the Chairman, Subcommittee on Risk Management, Research, and Specialty Crops, Committee on Agriculture, House of Representatives – Information on Prices of Genetically Modified Seeds in the United States and Argentina*, GAO/RCED/NSIAD-00-55. Washington DC: GAO, 12.

This article addresses some of these likely problems as they are now emerging. The article is structured as follows: an examination of the ways anticompetitive behaviours can restrict access to patent protected materials; an examination of limiting consumer uses through restrictive licensing practices; an examination of the competition effects of post patent regulatory approvals; an examination of the competition effects of post patent access to materials; an examination of customer and product shifting or switching and, finally, a discussion and some conclusions. The analysis in this chapter demonstrates that there are real competition concerns in the ways that patents have been deployed and used with GMOs. The various disputes, however, show that the parties themselves are generally able to find resolutions, albeit resolutions that may raise some concerns and are most likely to maintain the existing market power of key players in the markets for GMOs.

Accessing Patent Protected Materials

Perhaps the biggest patent and competition challenge posed by GMOs has been the lack of competition within and between rival GMO trait platforms. The Monsanto Corporation is really the only firm in the market that has a full suite of traits (see, for example, Table 4.1 showing the GMOs commercially available in Australia). This essentially means that Monsanto has a dominant position in the market for traits and their patents protect that competitive advantage. The following recent Monsanto and DuPont/Pioneer litigation highlights this competition problem.

Monsanto and DuPont/Pioneer entered into a non-exclusive licensing agreement whereby Monsanto licensed DuPont/Pioneer to use the events '40-3-2' and the 'NK603' involving both soybean and corn respectively.[9] The '40-3-2' and the 'NK603' events were the names given to the particular genetic modifications that resulted from inserting the novel DNA into the host plant. The re-issued US Patent 5,633,435 (from US Patent US RE 39,247E) with the priority date 13 September 1994 covers both events.[10] DuPont/Pioneer incorporated the '40-3-2' and 'NK603' events in their products and sold them, paying Monsanto a royalty.[11] After attempting to generate their own varieties to compete with

---

9 *Monsanto Company v. EI DuPont de Nemours & Co, Complaint (Redacted)*, 4:09-cv-00686 (Eastern District of Missouri, 2009), 3–4 ([7]–[9]); *Monsanto Company v. EI DuPont de Nemours & Co, Defendants Answers and Counterclaim (Redacted)*, 4:09-cv-00686 (Eastern District of Missouri, 2009), 2–3 ([7]–[9]).

10 *Monsanto Company v. EI DuPont de Nemours & Co, Complaint (Redacted)*, ibid., 3 ([7]). See also Gerard Barry, Ganesh Kishore, Stephen Padgette and William Stallings, *Glyphosate Tolerant 5-Enolpyruvylshikimate-3-Phosphate Synthase*, US Patent 5,633,435 (1997), Claim 1.

11 *Monsanto Company v. EI DuPont de Nemours & Co, Complaint (Redacted)*, ibid., 4 ([9]).

the '40-3-2' and 'NK603' events DuPont/Pioneer announced that they were combining their variety with these events.[12] As a consequence of combining the '40-3-2' and 'NK603' events with the DuPont/Pioneer events Monsanto alleged that DuPont/Pioneer had materially breached the licensing agreement[13] and infringed Monsanto's patent.[14] In response DuPont/Pioneer counterclaimed and, among other matters,[15] raised various anti-trusts defences.[16] The essence of their claim was:

> ... to arrest a new anticompetitive campaign by Monsanto designed to maintain and extend its unlawful monopolies into developing markets involving combinations ('stack') of input and output traits that confer multiple and more effective forms of herbicide tolerance or insect resistance or valuable end-use qualities. The chief means by which Monsanto has implemented this strategy has been either to directly deny competitors the ability to stack their traits with Monsanto's unlawfully acquired monopoly traits or to achieve the same result indirectly, by foreclosing competitors from access to the market by denying such stacking rights to independent seed companies. Monsanto has abused its unlawfully-acquired monopoly power to block competitors, thwart innovation and extract from farmers unjustified price increases of over 100 per cent in recent years.[17]

DuPont/Pioneer detailed four components to the scheme: stifling emerging competition to Monsanto's stacked multi-herbicide tolerant product (like competition from the DuPont/Pioneer Optimum GAT product) and shifting independent seed companies and farmers to another patented glyphosate resistant product (for example Roundup Ready 2, based on the 'MON89788' event); preventing independent seed companies from developing and selling products incorporating Monsanto's and others traits through restrictive licensing arrangements that exclude Monsanto's likely competitors and preventing stacking traits; entering anticompetitive agreements with trait developers that give Monsanto veto powers over sublicensing those traits to others, including likely competitors; and limiting access to germplasm into which traits can be placed and the traits that can be placed into germplasm.[18] DuPont/Pioneer identified the relevant markets

---

12   Ibid., 5 ([12]).

13   Ibid., 5 ([14]) and 16–22 ([76]–[103] and [110]–[122]).

14   Ibid., 5 ([14]) and 14–16 ([61]–[75]).

15   See *Monsanto Company v. EI DuPont de Nemours & Co, Defendants Answers and Counterclaim (Redacted)*, above n 9, 14–17 ([131]).

16   Ibid., 17 ([131]).

17   Ibid., 18–19 ([2]).

18   Ibid., 19–23 ([4]–[10]).

in the US[19] as the markets for 'herbicide-tolerant traits in soybeans',[20] 'herbicide-tolerant traits in corn.'[21] and 'stacked herbicide-tolerant and insect-resistant traits',[22] with Monsanto having the predominant market share (99.7 per cent for glyphosate resistant soybean,[23] 80.8 per cent for glyphosate resistant corn,[24] 74.9 per cent for corn rootworm resistance,[25] and 65.2 per cent for European corn borer resistance).[26] Moreover, DuPont/Pioneer suggested that:

> A small, but significant, non-transitory increase in licensing fees above the competitive levels for herbicide-tolerant traits in [the identified markets] would not cause seed companies or other customers to switch a significant enough quantity of purchases to another product so as to make the price increase unprofitable for a firm with monopoly power.[27]

DuPont/Pioneer also identified the emerging US markets for multiple modes of herbicide and insect resistance in corn,[28] stacked input and output traits in soybean,[29] and stacked input and output traits in corn[30] that Monsanto might also seek to restrict with its alleged anticompetitive activities.[31]

The other significant anticompetitive indications identified by DuPont/Pioneer were the significant barriers to entering the market against Monsanto[32] and the alleged demonstrated use of Monsanto's market power to increase seed prices.[33] DuPont/Pioneer identified two barriers to entering the market: first, time and costs of developing a new trait and second, resistance to getting that new trait to market because of Monsanto's exclusionary licensing of independent seed companies and farmers foreclosing their participation in taking up a new trait. On this second barrier, DuPont/Pioneer noted, '[t]he effect of such foreclosure will be to protect the monopoly power of Monsanto in both existing and emerging markets to the substantial detriment of competition and consumers'.[34]

---

19   Ibid., 33 ([55]).
20   Ibid., 26 ([21]).
21   Ibid., 26 ([27]).
22   Ibid., 28 ([32]).
23   Ibid., 26 ([24]).
24   Ibid., 27 ([29]).
25   Ibid., 28 ([34]).
26   Ibid., 28 ([34]).
27   Ibid., 26, 27 and 28–9 ([23], [28] and [36]).
28   Ibid., 29 ([40]).
29   Ibid., 30 ([44]).
30   Ibid., 32 ([50]).
31   See Ibid., 29–32 ([40]–[54]).
32   Ibid., 33–34 ([56]–[57]).
33   Ibid., 34 ([58]).
34   Ibid., 33–34 ([57]).

DuPont/Pioneer set out several factual examples of Monsanto's anticompetitive behaviour:

a. Product shifting – using market power to shift farmers from one product to another for the benefit of Monsanto and not the farmers – Monsanto's patent for the EPSPS gene is based on a claim for the isolated sequence.[35] This patent expires on 13 September 2014.[36] DuPont/Pioneer alleged that to extend its monopoly Monsanto is shifting (or switching) farmers to the Roundup Ready 2 varieties including entering into new licensing agreements with independent seed companies.[37] The Roundup Ready 2 has the same EPSPS gene except that the gene uses a different promoter and the construct of the promoter and the EPSPS gene is subject to a different patent that expires on 15 December 2020.[38] In effect the allegation is that Monsanto is extending its monopoly powers through intellectual property and contracting.[39] The contract is significant as, with its market power, Monsanto can demand that the independent seed companies convert their seeds lines to the Roundup Ready 2, and failing this Monsanto can terminate the license and require all the existing germplasm be destroyed.[40]

b. Restrictive licensing – using market power to exclude Monsanto's competitors – Monsanto is developing varieties of soybean stacked with resistance to glyphosate and other herbicides and insect resistance: 'SmartStax combines multiple insect-resistant traits operating by multiple modes of action and herbicide tolerant traits to multiple herbicides.'[41] DuPont/Pioneer assert that it co-developed the insect-resistant traits with Dow but requires Dow's permission to use the insect-resistant traits in seeds produced by independent seed companies.[42] DuPont/Pioneer also asserts that Dow has entered into an agreement with Monsanto to use the insect-resistant traits in the Monsanto SmartStax product and alleges that Monsanto's agreement includes an anticompetitive arrangement that

---

35 Barry *et al.*, above n 10, Claim 1.

36 This is 20 years after the priority date of 13 September 1994: ibid., 1.

37 *Monsanto Company v. EI DuPont de Nemours & Co, Defendants Answers and Counterclaim (Redacted)*, above n 10, 37–38 ([67]–[72]).

38 This is 20 years after the filing date of 15 December 2000: Fincher, K., Flasinski, S. and Wilkinson, J. 2003. *Plant Expression Constructs*, US Patent 6,660,911. Washington DC: USPTO, 1.

39 See *Monsanto Company v. EI DuPont de Nemours & Co, Defendants Answers and Counterclaim (Redacted)*, above n 10, 67–8 ([183]).

40 See ibid., 57–66 ([139]–[175]). See also Stumo, M. 2010. Anticompetitive tactics in ag biotech could stifle entrance of generic traits. *Drake Journal of Agricultural Law*, 15(1), 137–48, 138, 141–3 and the references therein.

41 *Monsanto Company v. EI DuPont de Nemours & Co, Defendants Answers and Counterclaim (Redacted)*, above n 10, 38 ([73]–[74]).

42 Ibid., 39 ([75]).

prohibits or penalises Dow for giving DuPont/Pioneer permission to sublicense the traits.[43] DuPont/Pioneer assert that this forecloses them from competing in the market for stacked herbicide and insect resistant products.[44]

This litigation was resolved with Monsanto and DuPont reaching a cross-license agreement allowing DuPont to stack Monsanto's Roundup Ready 2 with its materials.[45] The precise outcome of the settlement remains unclear. There are concerns that the cross-licensing arrangements might limit extending licensing rights to outside parties, creating more anti-competitive barriers. What this case starkly demonstrates is the broader scope of likely competition issues. Some of these are now addressed in more detail.

Limiting Consumer Uses Through Restrictive Licensing

The late expansion of patent law to include GMOs in 1980[46] and traditionally bred non-genetically modified plants in 2001[47] meant that plant and seed markets in agriculture had to rely on private ordering through contracts.[48] The result has been that plants and seeds are no longer sold but instead they are licensed, with the contract setting out the restrictive terms and conditions.[49] When combined with patents, the terms and conditions of such licenses attract the strict liability standard (including treble damages) and apply to innocent infringers through patent infringement rather than breach of contract. The decision of the US Supreme Court in *J.E.M. Ag Supply, Inc. v. Pioneer Hi-Bred International, Inc.* illustrates the kinds of accepted terms and conditions that restrict the purchaser:

> Pioneer sells its patented hybrid seeds under a limited label license that provides: 'License is granted solely to produce grain and/or forage' ... The license 'does not extend to the use of seed from such crop or the progeny thereof for

43   Ibid., 39 and 64–6 ([75] and [167]–[175]).

44   Ibid., 39 ([75]).

45   See Chao, B. and Gray, J. 2013. A $1 billion parable. *Denver University Law Review*, 90(1), 185–91.

46   *Diamond v. Chakrabarty*, above n 2, 305.

47   *J.E.M. Ag Supply, Inc. v. Pioneer Hi-Bred International, Inc.*, above n 4, 148.

48   See also Winston, E. 2012. A patent misperception. *Lewis and Clark Law Review*, 16(1), 289–335, 294–300 (arguing that the private ordering in agriculture is being allowed to expand on the patent bargain because of a misperception that patents are immune from antitrusts – '[i]t is anti-antitrusts to allow the abuse to continue'); Lawson, C. 2011. Juridifying the self-replicating to commodify the biological nature future: Patents, contracts and seeds, *Griffith Law Review*, 20(4), 851–82, 853–9 (arguing that the use of contracts and the recent interpretation of exhaustion provisions has expanded the reach of exclusivity).

49   See Winston, E. 2006. Why sell what you can license: Contracting around statutory protection of intellectual property. *George Mason Law Review*, 14(1), 93–133.

propagation or seed multiplication' … It strictly prohibits 'the use of such seed or the progeny thereof for propagation or seed multiplication or for production or development of a hybrid or different variety of seed'.[50]

How these contracts became so important and powerful for GMOs reflects an interesting choice made by the US courts in deciding the bounds of exhaustion. In three notorious cases farmers obtained GMOs in slightly different circumstances and then relied on the GMO traits without abiding by the patent holder's private ordering arrangements. The outcome has been that the courts have favoured the private ordering arrangements in crafting their resolutions to the disputes.

In *Monsanto Company v. McFarling* in the US Court of Appeals McFarling had purchased some of Monsanto's patent protected Roundup Ready soybean seeds under a 'Technology Agreement' and paid a license fee.[51] The terms of the 'Technology Agreement' included that the seeds would be used 'for planting a commercial crop only in a single season' and that the licensee would not 'save any crop produced from this seed for replanting, or supply saved seeds to anyone for replanting'.[52] McFarling did save some seeds from his crop and then used those seeds the following cropping season and repeated this activity in subsequent cropping seasons.[53] McFarling did not pay Monsanto any fees for the saved seeds and their subsequent use in cropping.[54]

McFarling argued, in part, that the 'Technology Agreement' required him to purchase new seed each season from Monsanto and that this unreasonably broadened the patent grant as he was well able to produce his own seed from his existing purchases of Monsanto's patent protected seeds.[55] The majority, Judges Newman and Bryson, rejected this proposition finding that other seeds were available to McFarling and that Monsanto's 'Technology Agreement' applied to 'the purchased seed for the purpose of growing crops and not for the purpose of producing new seed'.[56] Thus:

---

50   *J.E.M. Ag Supply, Inc. v. Pioneer Hi-Bred International, Inc.*, above n 4, 128.

51   *Monsanto Company v. McFarling*, 302 F.3d 1291, 1293 (2002).

52   Ibid., 1293.

53   Ibid.

54   Ibid.

55   Ibid., 1297.

56   Ibid., 1298. This patent has been extensively litigated, leading to finding that planting seeds containing the invented sequences is an infringement: see *Monsanto Company v. David*, 516 F.3d 1009 (2008); *Monsanto Company v. Parr*, 545 F.Supp.2d 836 (2008); *Monsanto Company v. Vanderhoof*, 2007 WL 1240258 (2007); *Monsanto Company v. Strickland*, 2007 WL 3046700 (2007); *Monsanto Company v. Good*, 2004 WL 1664013 (2003); *Monsanto Company v. Trantham*, 156 F.Supp.2d (2001); *Monsanto Company v. Dawson*, 2000 WL 33953542 (2000); and so on.

A purchaser's desire to buy a superior product does not require benevolent behaviour by the purveyor of the superior product. Nor does an inventor of new technology violate the antitrust laws merely because its patented product is favoured by consumers.[57]

McFarling's challenge was also, in part, that the 'Technology Agreement' violated the doctrines of patent exhaustion (or first-sale), and the proposition that 'when a patented product has been sold the purchaser acquires "the right to use and sell it, and the authorized sale of an article which is capable of use only in practicing the patent is a relinquishment of the patent monopoly with respect to the article sold".'[58] As a consequence McFarling asserted that Monsanto's sale of the patent protected seeds to McFarling exhausted the 'exclusive rights' over the seeds and products and that this could not be restricted by the 'Technology Agreement'.[59] The majority accepted Monsanto's response that the seeds were not sold. Under the 'Technology Agreement' Monsanto sold use rights but not the rights to construct new seeds.[60] The majority relied on the earlier Federal Circuit decision in *Mallinckrodt, Inc. v. Medipart, Inc.* for the proposition that 'use of a patented product "in violation of a valid restriction may be remedied under the patent law, provided that no other law prevents enforcement of the patent".'[61] The conclusion was that Monsanto's authorised sale of the patent protected seeds with the 'Technology Agreement' restriction did not exhaust the 'exclusive rights' but instead was a limited permission to use the seeds for specific purposes according to contract laws that might be remedied under patent law.[62]

In *Monsanto Company v. Scruggs* in the US Court of Appeals, Scruggs bought seeds containing Monsanto's patent protected invention (in this case also including insect resistance in addition to herbicide resistance). Scruggs planted his purchased seeds, saved seeds from the harvest and planted subsequent generations.[63] The important difference was that Scruggs did not sign the 'Technology Agreement'.[64] The seed company from which Scruggs bought the seed was not Monsanto but rather a company licensed to sell the Monsanto seeds, including the requirement to sell them with the signed 'Technology Agreement'.[65] Monsanto filed an infringement suit and was awarded a permanent injunction.[66] Scruggs lodged an appeal arguing, in part, that Monsanto's patent was exhausted because Scruggs had

---

57 Ibid., 1298.
58 Ibid., citing *United States v. Univis Lens Co*, 316 US 241, 249 (1942).
59 *Monsanto Company v. McFarling*, Ibid., 1298.
60 Ibid., 1298–9.
61 Ibid., 1298 citing *Mallinckrodt, Inc. v. Medipart*, Inc., 976 F.2d 700, 701 (1992).
62 Ibid., 1299–1300.
63 *Monsanto Company v. Scruggs*, 459 F.3d 1328, 1333 (2006).
64 Ibid.
65 Ibid.
66 Ibid.

purchased the Monsanto seeds in an unrestricted sale.[67] Scruggs' argument was that because his purchase was an unrestricted sale the patent exhausted on sale.[68] Despite this argument, Circuit Judge Mayer, with Circuit Judge Dyk concurring on this issue,[69] rejected his proposition:

> The doctrine of patent exhaustion is inapplicable in this case. There was no unrestricted sale because the use of the seeds by seed growers was conditioned on obtaining a license from Monsanto. Furthermore, the '"first sale" doctrine of exhaustion of the patent right is not implicated, as the new seeds grown from the original batch had never been sold' ... Without the actual sale of the second generation seed to Scruggs, there can be no patent exhaustion. The fact that a patented technology can replicate itself does not give a purchaser the right to use replicated copies of the technology. Applying the first sale doctrine to subsequent generations of self-replicating technology would eviscerate the rights of the patent holder.[70]

More recently in *Monsanto Company v. Bowman* a farmer bought seeds from a grain elevator (where seeds are delivered as a commodity after harvest for transport to markets) for the purposes of planting and harvesting a second season crop.[71] Specifically Bowman conformed with Monsanto's licensing arrangements for the first season crop, but then planted a second season crop using mixed seeds, some of which contained Monsanto's patented technology that he had obtained from a grain elevator without the 'Technology Agreement'.[72] Bowman also saved some of the seed from the second season crop for planting in subsequent seasons' second season crop.[73] Monsanto alleged patent infringement.[74] Bowman argued:

> that when the soybeans from a licensed Roundup Ready crop are harvested and sold to a grain elevator or dealer, they are sold without restriction, mixed with all other soybean crops in from the area and, therefore, when purchased and used by farmers to plant as seed (commodity soybeans) for another crop, they are not protected by patent.[75]

---

67   Ibid., 1335.
68   Ibid., 1335–6 citing *Mallinckrodt, Inc. v. Medipart*, Inc., 976 F.2d 700, 701 (1992). Notably the decision also refers to the decision in *LG Electronics, Inc. v. Bizcom Electronics, Inc.* 453 F. 3d 1364 (2006) that was expressly overruled on appeal in the Supreme Court in *Quanta Computer, Inc. v. LG Electrics, Inc.*, 553 US 617, 638 (2008).
69   Ibid., 1342.
70   Ibid., 1336 citing *Monsanto v. McFarling*, above n 52, 1299.
71   *Monsanto Company v. Bowman* 686 F.Supp.2d 834, 835 (2009).
72   Ibid., 835–6.
73   Ibid., 836.
74   Ibid., 836.
75   Ibid.

The District Court followed the earlier decisions in *McFarling* and *Scruggs* and found that Bowman had infringed Monsanto's patent because the seeds from the grain elevator containing the patented invention were expressly excluded by agreement from being sold for planting – 'No unconditional sale of the Roundup Ready trait occurred because the farmers could not convey to the grain dealers what they did not possess themselves.'[76] The District Court concluded:

> Unless Monsanto receives the patent protection it is trying to enforce in this case … there would be nothing stopping all farmers from buying commodity soybeans for planting … thereby allowing such farmers to receive the benefit of the Roundup Ready genetic modification without compensating Monsanto for its research and development work.[77]

The US Court of Appeals affirmed this decision, reasoning that patent exhaustion did not occur because Bowman had 'created a newly infringing article'.[78] The decision was again affirmed in the US Supreme Court.[79] Essentially the dilemma for the court was that patent exhaustion allows a purchaser in an unrestricted sale to use the invention without restriction but not to remake or replicate the invention – 'That is because the patent holder has "received his reward" only for the actual article sold, and not for subsequent recreations of it.'[80] Seeds, however, contain their own means of reproduction and by using the seed as intended the seed reproduces itself as a crop – 'he is merely using them in the normal way farmers do'.[81] The court concluded:

> Unfortunately for Bowman, that principle decides this case against him. Under the patent exhaustion doctrine, Bowman could resell the patented soybeans he purchased from the grain elevator; so too he could consume the beans himself or feed them to his animals. Monsanto, although the patent holder, would have no business interfering in those uses of Roundup Ready beans. But the exhaustion doctrine does not enable Bowman to make additional patented soybeans without Monsanto's permission (either express or implied). And that is precisely what Bowman did. He took the soybeans he purchased home; planted them in his fields at the time he thought best; applied glyphosate to kill weeds (as well as any soy plants lacking the Roundup Ready trait); and finally harvested more (many more) beans than he started with. That is how 'to

---

76 Ibid., 839 citing *Monsanto Company v. Scruggs*, 459 F.3d 1328, 1336 (2006).
77 *Monsanto Company v. Bowman*, 657 F.3d 1341, 1348 (2011).
78 Ibid.
79 *Bowman v. Monsanto Company*, 133 S.Ct. 1761 (2013).
80 See *Bowman v. Monsanto Company*, ibid., 1766 citing *United States v. Univis Lens Co.*, 316 U.S. 241, 251 (1942).
81 Ibid.

"make" a new product', to use Bowman's words, when the original product is a seed ... Because Bowman thus reproduced Monsanto's patented invention, the exhaustion doctrine does not protect him.[82]

The Supreme Court justified this perspective, arguing that if there were exhaustion on first sale then Monsanto would only ever '"receive its reward" for the first seeds it sells'[83] and:

in short order, other seed companies could reproduce the product and market it to growers, thus depriving Monsanto of its monopoly. And farmers themselves need only buy the seed once, whether from Monsanto, a competitor, or (as here) a grain elevator. The grower could multiply his initial purchase, and then multiply that new creation, ad infinitum – each time profiting from the patented seed without compensating its inventor. Bowman's late-season plantings offer a prime illustration. After buying beans for a single harvest, Bowman saved enough seed each year to reduce or eliminate the need for additional purchases. Monsanto still held its patent, but received no gain from Bowman's annual production and sale of Roundup Ready soybeans.[84]

While the court was careful to say that this decision was not the final statement on self-replicating technologies,[85] the effect is to favour Monsanto's private ordering according to its particular terms and conditions. In effect the patent issue of exhaustion has been decided in favour of the patent holder and becomes a mere question about making the invention.

Post-patent Regulatory Approvals

Most nations require GMOs to be approved by a governmental agency for growing and selling. These nations have either adopted specific regulations directed to GMOs or have applied their existing general regulations (with some amendments) to accommodate GMOs. For example, in the US the existing general regulations are applied by the US Department of Agriculture (USDA), the Food and Drug Administration (FDA) and the Environmental Protection Agency (EPA). The particular agency responsible for overseeing the regulation of any given GMO depends on its intended use.[86] The FDA regulates food uses through a voluntary consultation process, the EPA regulates potential

---

82   Ibid., 1766–7.

83   Ibid., 1767 citing *United States v. Univis Lens Co.*, 316 U.S. 241, 251 (1942).

84   *Bowman v. Monsanto Company*, ibid., 1767.

85   Ibid., 1769. See also Sheff, J. 2013. Self-replicating technologies. *Stanford Technology Law Review*, 16(2), 229–56.

86   See *Coordinated Framework for Regulation of Biotechnology*, 51 Fed. Reg. 23,302.

pesticide harms to the environment through the *Federal Insecticide, Fungicide and Rodenticide Act*,[87] and the USDA regulates plant pests and noxious weeds through the *Plant Protection Act*,[88] albeit that other laws may also have some application.[89] The effect of governmental regulation is that GMOs require regulatory approval (generally in the form of a licence) to be grown, distributed, sold or consumed in the market.[90] The regulatory approvals are generally time limited and expire, and require some kind of involvement of the registrant to provide testing data held by the original applicant. These kinds of schemes apply in most markets around the world, meaning that to import and use GMOs requires regulatory approvals.[91]

The requirements for regulatory approval raise competition issues about who will maintain the regulatory approval following the patent term where the patent holder (and their licensees or assignees) ceases to maintain that approval. Theoretically when a patent expires the subject matter of the patent becomes available to the broader market and generic competition is then possible (increasing access and lowering the prices). Where a patent protected product or process also requires other regulatory approval then that approval may become a barrier to competition. These might be considered as two separate competition problems in: (1) regulatory approval in the domestic market for GMOs and their products and (2) regulatory approval in export markets and destinations for GMOs and their products. A patent holder may choose not to maintain regulatory approval with the effect that

---

87   7 U.S.C. 6. *The Federal Insecticide, Fungicide and Rodenticide Act* gives the EPA authority to limit the sale and distribution of pesticides: 7 U.S.C. § 136 et seq.; 40 C.F.R. §§ 152 and 174.

88   7 U.S.C. 104. *The Plant Protection Act* gives the USDA authority to 'prohibit ore restrict ... any plant ... if [the Secretary of Agriculture] determines that the prohibition or restriction is necessary to prevent ... the dissemination of a plant pest or noxious weed within the United States': 7 U.S.C. § 7712(a). The term 'plant pest' has been broadly interpreted to give the USDA authority (delegated to the Animal and Plant Health Inspection Service): see 7 C.F.R. § 340.

89   See, for examples: FDA – *Food, Drug and Cosmetics Act*, 21 U.S.C. 9; Public Health Service Act, 42 U.S.C. 6A; EPA – *National Environmental Protection Act*, 42 U.S.C. 55; *Toxic Substances Control Act*, 15 U.S.C. 53; *Food, Drug and Cosmetics Act*, 21 U.S.C. 9; USDA – *Virus Serum Toxin Act*, 21 U.S.C. 5; *Meat Inspection Act*, 21 U.S.C. 12; *Poultry Products Inspection Act*, 21 U.S.C. 10; *Egg Products Inspection Act*, 21 U.S.C. § 15 and so on.

90   For an overview of this complex regulatory thicket in the US see McHughen, A. 2006. *Plant Genetic Engineering and Regulation in the United States*, Agricultural Biotechnology in California Series, Publication 8179. [Online]. Available at: http://anrcatalog.ucdavis.edu/pdf/8179.pdf [accessed 3 February 2014].

91   For a general, partisan perspective see Conko, G. 2012. *Is There a Future for Generic Biotech Crops? Regulatory Reform is Needed for a Viable Post-Patent Industry*, Issue Analysis 2012 No 7, 4–10. [Online]. Available at: http://cei.org/sites/default/files/Greg%20Conko%20-%20Is%20There%20a%20Future%20for%20Generic%20Biotech%20Crops.pdf [accessed 3 February 2014].

patent term expiry removes the GMO from the market because there is no longer a regulatory approval to grow, distribute, sell and consume the GMO in the market.[92]

That this is likely to be a problem was well illustrated in the *In re Genetically Modified Rice Litigation*.[93] In this case Bayer CropScience was advised that the rice variety LLRice601 (which was unapproved in the US and Europe and still experimental) was found in the 2005 long grain rice harvest. Damages were awarded against Bayer, partly recognising the economic impact of failing to have regulatory approval in export markets:

> Numerous Bayer documents show that Bayer knew the LL601 had to be kept isolated, could not enter the food chain, and could not enter foreign markets. Bayer employees referred to Europe's 'zero tolerance' policies. They discussed the effects of such an event on the market for rice, recognizing that there could be serious economic impacts. One 2000 memorandum even correctly forecast that if GM rice was found to have spread to conventional varieties, 'We could make any national newscast ... and the rice industry would be quite affected to say the least ...'[94]

The imminent expiry of the Monsanto patent over Roundup Ready 1 on 13 September 2014[95] (and another 22 biotech traits over the next decade)[96] demonstrated the likely competition problems and how they have been addressed in anticipation of that event. The particular concern here is that, now Monsanto has regulatory approval for its patent protected Roundup Ready 2, the company will have little interest in maintaining regulatory approval for Roundup Ready 1 when it comes off patent. The immediate concern about Monsanto Roundup Ready 1 has probably been put off, at least for soybeans, as Monsanto has committed to 'continue global regulatory approvals through 2021 for RR 1 soybeans'.[97] To address this concern five major seed companies have agreed to a binding legal contract covering the expiry of single gene patents. The *Generic Event Marketability and Access*

---

92   See generally McEowen, R. 2011. *Expiration of Biotech Crop Patents – Issues For Growers*, Iowa State University Center for Agricultural Law and Taxation Issue Brief, 8 April 2011. [Online]. Available at: http://www.calt.iastate.edu/briefs/CALT%20Legal%20 Brief%20-%20Expiration%20of%20Biotech%20Crop%20Patents%20-%20Issues%20 for%20Growers.pdf [accessed 3 February 2014].

93   In *re Genetically Modified Rice Litigation* 666 F.Supp.2d 1004, 1014–15 (Eastern District of Missouri, 2009).

94   Ibid., 1031.

95   This is 20 years after the priority date of 13 September 1994: Barry *et al.*, above n 10, 1.

96   See Conko, above n 91, 1.

97   See Jones, P. 2012. *Both Expiring and Healthy Patents Breed Challenges for Ag Biotech*, ISB News Report. [Online]. Available at: http://www.isb.vt.edu/news/2012/feb/ jones-1.pdf [accessed 3 February 2014]. See also Redick, T. and Hawker, N. 2010. Legal issues arising from generic biotech crops. *Aglaw Update*, 27(3), 2–6.

Agreement[98] and the *Data Use and Compensation Agreement* are the relevant agreements. The *Generic Event Marketability and Access Agreement* sets out agreements about accessing patent protected products and any regulatory data and authorisations. The *Data Use and Compensation Agreement* sets out agreement to encourage signatories to share or transition responsibility for regulatory maintenance, stewardship and liability.[99]

In overview the *Generic Event Marketability and Access Agreement* provides that the signatories agree:

To 'negotiate in good faith to grant access' to the single event patent protected product and any regulatory data and authorisations.[100]

To make the single event patent protected product available to all signatories after patent expiry in non-proprietary germplasm.[101]

To provide notice to the other signatories of the imminent expiry of a single event patent.[102]

To indicate whether the regulatory approval will be maintained, shared or discontinued by the patent holder.[103]

To establish an administration entity to operate and administer the agreement.[104]

Under the *Generic Event Marketability and Access Agreement* so far there have been two notifications of patent expiry: the Monsanto patent for event MON 810 Corn and the 40-3-2 Soybean.[105] Monsanto has committed to maintain the authorisations necessary for cultivation and sale of seed products and grain in the US and 'to permit undisrupted trade' in materials containing these events.[106]

While the *Generic Event Marketability and Access Agreement* and the *Data Use and Compensation Agreement* are an attempt to address the competition problems of regulatory approval stifling access and use of post-patent products,

---

98   *Generic Event Marketability and Access Agreement.* [Online]. Available at: http://www.agaccord.org/include/gemaa_firstamendedMay9.pdf [accessed 3 February 2014].

99   Ibid., cl 6.

100   Ibid., cl 4.

101   Ibid., cl 5.

102   Ibid., cl 6.

103   Ibid., cls 6, 7 (maintain), 8 (share) and 9 (discontinue).

104   Ibid., cl 26 and Appendix A.

105   Generic Event Marketability and Access Agreement, *Notice of Patent Expiry – Mon 810 Corn* and *Notice of Patent Expiry – 40-3-2 Soybean*. [Online]. Available at: http://www.agaccord.org/?p=notices [accessed 3 February 2014].

106   Ibid.

it is still too early to determine whether this will be sufficient. Notably the agreements really only deal with the other signatories and, at this stage, the process of dealing with shared or discontinued regulatory approval and multiple event traits (stacked traits) is uncertain. The positive outcome for competition, though, is that Monsanto has committed to maintain existing regulatory approval for its two expiring traits that should enable a generic adoption of the patented traits to be exploited after patent term expiry.[107]

Post-patent Access to Materials

A closely related issue is whether at the expiration of a patent the protected product or process becomes available for generic competition. Underpinning the justification of intellectual property schemes is a period of exclusivity, after which the intellectual property protected material becomes available to the broader public. The *Agreement of Trade-related Aspects of Intellectual Property* (TRIPS) minimum standard provides that the period of exclusivity is 20 years from the filing date.[108] The patent application itself requires a description that discloses the invention in a way that a person skilled in the art might carry out the invention.[109] The contents of the description include a specification setting out details about the invention and how it might be performed.[110] For complex materials, such as DNA constructs, plants, seeds and so on, depositing the materials with a suitable repository can satisfy the description requirements.[111] This overcomes the problem of dealing with materials that might otherwise defy description (such as plants, cells, seeds and the like). The depositor (the patent applicant and/or holder) however, only has obligations to ensure the deposit remains viable during the term of the patent and there are no obligations on the repository either to maintain the materials or to make them available after the patent term expires. Similarly, TRIPS and other intellectual property agreements do not require that materials deposited to satisfy the description requirements are made available after the patent term expires. The result is that patented materials may not be accessible after the patent term expires.

107   See Sauer, K. 2014. *Roundup Ready® Soybean Post-Patent Regulatory Commitment Extended through 2021*. [Online]. Available at: at http://www.monsanto.com/newsviews/pages/roundup-ready-soybean-post-patent-commitment-extended-through-2021.aspx [accessed 3 February 2014].

108   *Agreement of Trade-related Aspects of Intellectual Property*, Article 33. See also *Canada – Term of Patent Protection* (2000) WT/DS170/AB/R, [95].

109   *Agreement of Trade-related Aspects of Intellectual Property*, Article 29(1).

110   See *Patent Cooperation Treaty*, Article 5; *Regulations under the Patent Cooperation Treaty*, Article 5.

111   *Budapest Treaty on the International Recognition of the Deposit of Micro-organisms for the Purposes of Patent Procedure*, Article 3.

The *Generic Event Marketability and Access Agreement* addresses this issue by providing that the signatories agree to make the single event patent protected product available to all signatories after patent expiry in the form of non-proprietary germplasm.[112] While this is a solution for the signatories, this is not a solution for non-signatories and their disadvantage may be exacerbated by existing practices where patent protected products are made available under a license that restricts seed saving and compels destruction of existing seed stocks.

Customer/Product Shifting or Switching

Patent holders are often motivated to maintain their intellectual property exclusivity by shifting or switching the market (also called product hopping and ever-greening) in favor of their intellectual property protected product. This has been a well-documented phenomenon in the pharmaceutical industry where, just before patent expiry, the patent holder adopts counter measures to extend their favoured position by shifting or switching customers to another substituted product that is patent protected.[113] Two recent examples demonstrate the kinds of competition issues that arise when a patent holder actively seeks to secure an advantage through switching or shifting. Both involve Monsanto attempting to avoid generic competition by moving customers: (1) from a less robustly protected product to a more robustly protected product and (2) from an unprotected product to a protected product. In the latter case Monsanto has made a number of concessions, almost certainly to avoid competition (antitrust) concerns.

The first example involved alleged attempts by Monsanto to shift or switch customers from the GA21 event to the NK603 event. In essence Monsanto's patent entitlements over the GA21 event were successfully challenged so that Monsanto no longer retained exclusivity.[114] Monsanto did, however, have exclusive patent entitlements over the NK603 event.[115] In a patent spat between Monsanto and Syngenta concerning the event GA21,[116] Syngenta Seed responded in the US District Court in Delaware, alleging Monsanto:

---

112   *Generic Event Marketability and Access Agreement*, above n 98, cl 5.

113   For examples, see Westin, J. 2011. Product switching in the pharmaceutical sector – An abuse or legitimate commercial consideration. *European Competition Law Review*, 32(12), 595–601.

114   See *Rhone-Poulenc Agro, S.A. v. Dekalb Genetics Corporation*, 272 F.3d 1335 (2001).

115   See Behr, C., Heck, G., Hironaka, C. and You, J. 2004. *Corn Plants Comprising Event PV-ZMGT32 (nk603)*, US Patent 6,825,400. Washington DC: USPTO.

116   See Jones, P. 2005. *Patent Challenges to AgBiotech Technologies in 2004*, ISB News Report. [Online]. Available at: http://www.isb.vt.edu/news/2005/artspdf/feb0504. pdf [accessed 3 February 2014]; Market Scope Europe Ltd. 2004. *European News And Markets – Crop Protection Monthly*, Issue No 176, 3. [Online]. Available at: www.crop-protection-monthly.co.uk/Archives/CPMJuly2004.doc [accessed 3 February 2014].

to have: (1) bundled commercial incentives across several products sold to corn growers to create a barrier to plaintiff and other competitors ... (2) enforced exclusive dealing contracts to prevent plaintiff or other competitors from entering markets ... (3) filed the 'baseless' [patent infringement litigation] against plaintiff ... (4) filed separate 'baseless' patent case in Illinois concerning two other glyphosate-tolerant trait patents ... (5) misrepresented plaintiff's ability to commercialize glyphosate-tolerant traits to discourage seed companies from dealing with plaintiff ... (6) demanded destruction of all GA21 production to impair plaintiff's entry into the glyphosate-tolerant traits market ... (7) intimidated seed companies not to do business with plaintiff ... and (8) denied plaintiff access to foundation seeds and pressured foundation seed companies not to deal with plaintiff.[117]

Syngenta and Monsanto eventually settled all their litigation and entered into a licensing arrangement.[118]

The second example is Monsanto's shifting or switching of customers from the patent expiring Roundup Ready 1 to the patent protected Roundup Ready 2. Roundup Ready 2 has the same glyphosate resistance gene linked to a different promoter to activate that gene and the patent expires on 15 December 2020.[119] Monsanto now appears to be slowly shifting its existing uses of Roundup Ready 1 to Roundup Ready 2 and requiring its licensed independent seed sellers to switch to Roundup Ready 2, according to their contractual arrangements:

Notwithstanding that [Roundup Ready] and [Roundup Ready 2] are virtually identical, it is widely known in the seed industry that Monsanto recently informed independent seed companies ... that they must begin to convert all of their soybean seed lines from [Roundup Ready] to [Roundup Ready 2] within three years if they wish to continue licensing [Roundup Ready]. Otherwise, Monsanto will terminate the [independent seed companies'] license for [Roundup Ready] soybeans and require the [independent seed company] to destroy all of its [Roundup Ready] soybean germplasm. Because many farmers today will not purchase soybean seeds without a glyphosate-tolerant trait, [independent seed companies] face the prospect of losing their Monsanto license and being driven from the soybean seed market unless they agree to switch to [Roundup Ready 2] completely.[120]

---

117   *Syngenta Seed, Inc. v. Monsanto Company*, 2005 WL 678855, 2 (D.Del.).

118   See Zuhn, D. 2008. *Monsanto and Syngenta Settle All Litigation Between the Companies*. [Online]. Available at: http://www.patentdocs.org/2008/05/monsanto-and-sy.html [accessed 3 February 2014].

119   This is 20 years after the filing date of 15 December 2000: Fincher et al., above n 38, 1.

120   Stumo, above n 40, 141–2.

If Monsanto can shift to Roundup Ready 2 then it will be able to avoid generic competition and, in particular, competition from farmers saving their seeds from earlier crops to grow again. Switching to Roundup Ready 2 also allows Monsanto to maintain control over growers through the threat of patent infringement rather than just through contracts and protections of IP exclusivity.

Discussion and Conclusions

Central to appreciating the patent and competition issues for GMOs is an appreciation of the specific and limited nature of genetic traits and their limited substitutability. For example, three basic strategies have been used to develop glyphosate tolerance in crop species: over-expression of EPSP synthase, expression of an insensitive EPSP synthase and detoxification of the N-phosphonomethylglycine molecule.[121] Neither over-expression of EPSP synthase nor detoxification of the N-phosphonomethylglycine molecule has proved viable because of the secondary effects of overexpression and insufficient tolerance to glyphosate.[122] This means that over-expression of EPSP synthase is really the only viable alternative with no other substitutes. This is because of an inherent limitation in biology: common evolutionary ancestry means that the same biological mechanisms are broadly distributed across living organisms. Nature has used the same tricks many times over. As a generalisation, therefore, GMOs pose particular competition problems because their traits are generally not substitutable. This raises particular competition problems as patents deliver exclusivity to the rights holder that can be expertly protected and exploited through careful contracting arrangements. The exploitation of GMOs is an exemplar of how patents can best be exploited, and directly challenges the boundary between patent laws and competition laws.

The analysis in this chapter posits that there is a problem of market concentration in agriculture that is being enabled by intellectual property, and particularly patents, over agriculturally significant traits. The main competition problems might be characterised as:

A limited number of technology platforms – only one firm, Monsanto, has a full suite of traits that can be stacked into a final product. They also own many of the key transformational technologies, including the agrobacterium co-transformation method and the anti-biotic genes under the control of a plant promoter. Moreover, there is a limited stock of elite germplasm that is increasingly owned by a private sector that provides limited access to competitors.

---

121   Dill, G. 2005. Glyphosate-resistant crops: History, status and future. *Pest Management Science*, 61(3), 219–24, 219 and the references therein.

122   See *Monsanto Company v. EI DuPont de Nemours & Co, Complaint (Redacted)*, above n 10, 13–14 ([57]–[58]).

A limited rivalry between platforms – Monsanto has not allowed its traits to be broadly adopted by competitors. One result of this restricted access is a concentration of the market share across a range of crops in Monsanto's hands. In effect, a single player has managed to integrate its traits across platforms limiting the potential of competition between platforms.

Mergers and acquisitions – competing firms are often acquired by the bigger players so that their technology and knowledge is being concentrated into a few very large firms.

High barriers to entry – the relatively high costs of research and development that are required to participate in this market act as a barrier to entry for new participants. The merger and acquisition by bigger players of any potential competitors exacerbates this problem.

Vertical integration – firms such as Monsanto control downstream players by forcing them to sign restrictive agreements. In the case of agriculture this includes restrictive agreements with germplasm providers, seed sellers and farmers.

The consequences of these competition problems are likely to be lower quality innovation occurring at a reduced frequency. The imperative on maintaining existing power shifts innovative focus in the sector away from producing new inventions and towards protecting existing inventions and existing market arrangements, higher seed prices and fewer product choices. The analysis in this chapter demonstrates that these are real competition concerns, arising from the ways that patents have been deployed and used with GMOs. The various disputes between GMO developers and sellers show that the parties themselves are generally able to find resolutions, albeit that those resolutions may raise competition concerns and maintain the existing market power of the key players in the various markets for GMOs. Most importantly, the likely consequence of reduced competition is a price squeeze on farmers as they offset higher seed prices against fairly stable global commodity market prices. In making this trade-off farmers are accepting that they are the weakest bargainers in the chain and will continue to receive a reduced share of the profits created by GMOs.

Table 4.1   Commercial uncontrolled released GMOs in Australia up to 2013. This table is compiled from the licenses granted for the uncontrolled release of GMOs into the environment by the Gene Technology Regulator under the *Gene Technology Act 2000* (Cth)

Plant	Owner	Branding	Event	Description
Cotton (Gossypium hirsutum L.)	Monsanto Australia Ltd	INGARD (Bollgard)	MON531	Insecticide – cry1Ac gene from Bacillus thuringiensis
		Roundup Ready	MON1445	Glyphosate tolerance –cp4 epsps gene from Agrobacterium species CP4
		Roundup Ready/ INGARD	MON1445 x MON531	Insecticide – cry1Ac gene from Bacillus thuringiensis Glyphosate tolerance –cp4 epsps gene from Agrobacterium species CP4
		Bollgard II	MON15985	Insecticide – cry1Ac and cry2Ab genes from Bacillus thuringiensis
		Bollgard II/ Roundup Ready	MON15985 x MON1445	Insecticide – cry1Ac and cry2Ab genes from Bacillus thuringiensis Glyphosate tolerance –cp4 epsps gene from Agrobacterium species CP4
		Roundup Ready Flex	MON88913	Glyphosate tolerance – two modified cp4 epsps genes from Agrobacterium species CP4
		Roundup Ready Flex/ Bollgard II	MON88913 x MON15985	Insecticide – cry1Ac and cry2Ab genes from Bacillus thuringiensis Glyphosate tolerance – two cp4 epsps genes from Agrobacterium species CP4
	Bayer CropScience Pty Ltd	Liberty Link	LLCotton25	Glufosinate ammonium tolerance – bar gene from Streptomyces hygroscopicus
	Dow AgroSciences Australia Pty Ltd	WideStrike	281-24-236 x 3006-21-23	Insecticide – synthetic cry1Ac(synpro) and cry1F(synpro) genes and from Bacillus thuringiensis
Canola (Brassica napus L.)	Monsanto Australia Ltd	Roundup Ready	GT73	Glyphosate tolerance – cp4 epsps gene from Agrobacterium species CP4
	Bayer CropScience Pty Ltd	InVigor	MS8 RF3	Glufosinate ammonium tolerance – bar gene from Streptomyces hygroscopicus Fertility – barnase (male sterility; MS8) and barstar (fertility restorer; RF3) genes Bacillus amyloliquefaciens

Table 4.1   *Continued*

Plant	Owner	Branding	Event	Description
Canola (Brassica napus L.)	Bayer CropScience Pty Ltd	InVigor	T45	Glufosinate ammonium tolerance – pat gene from Streptomyces hygroscopicus
			Topas 19/2	Glufosinate ammonium tolerance – pat gene from Streptomyces hygroscopicus
			MS1	Glufosinate ammonium tolerance – bar gene from Streptomyces hygroscopicus Fertility – barnase (male sterility) gene from Bacillus amyloliquefaciens
			RF1 and RF2	Glufosinate ammonium tolerance – bar gene from Streptomyces hygroscopicus Fertility – barstar (fertility restorer) gene from Bacillus amyloliquefaciens
			MS8	Glufosinate ammonium tolerance – bar gene from Streptomyces hygroscopicus Fertility – barnase (male sterility) gene from Bacillus amyloliquefaciens
			RF3	Glufosinate ammonium tolerance – bar gene from Streptomyces hygroscopicus Fertility – barstar (fertility restorer) gene from Bacillus amyloliquefaciens
		InVigor x Roundup Ready	MS8/RF3 x GT73	Glufosinate ammonium tolerance – bar gene from Streptomyces hygroscopicus Fertility – barnase (male sterility; MS8) and barstar (fertility restorer; RF3) genes from Bacillus amyloliquefaciens Glyphosate tolerance –cp4 epsps gene from Agrobacterium species CP4 and the goxv247 gene from Ochrobactrum anthropi
Rose (Rosa X hybrida)	Florigene Pty Ltd	GM Hybrid Tea	WKS82/130-4-1	Purple/blue flower colour – Flavonoid 3'5'-hydroxylase gene from Viola x wittrockiana and the Anthocyanin 5-acyltransferase gene from Torenia x hybrida

Chapter 5

Regulating for Traditional Innovation in Agricultural Organisms

Karinne Ludlow

Introduction

Whilst any particular innovation may be new, innovation more broadly is not. In the context of agriculture, innovation through genetic change is fundamental. The genetic makeup of agricultural plants and animals used in any particular geographic location may originally have been endemic or imported,[1] but it will almost certainly have been changed through human direction and innovation. New genetic makeups, or genotypes, are developed not only in laboratories but also by small-scale farmers and indigenous and local communities. Indeed ongoing innovation by all groups is necessary for the continued productivity and intensification required to feed the future world.[2] The contribution of small-scale farmers and local and indigenous communities is often expressed as being based on conserving traditional varieties; their role in the innovative process by which genetic resources are continually refined and developed, called 'traditional innovation' in this chapter, is frequently overlooked.[3] Nevertheless, the ability of

---

1 For example, a review of Australia's biosecurity observed that 'it is often forgotten that almost all the crops and animals (and much of the pastures) forming the basis of Australian agriculture were initially imported into the country ... Moreover, researchers and producers alike are constantly scouring the world for improved genetic material as part of the relentless challenge of enhancing international competitiveness.' Commonwealth of Australia. 2008. *One biosecurity. A working partnership. The independent review of Australia's quarantine and biosecurity arrangements report to the Australian Government* (Beale Report), XVI. [Online]. Available at: http://www.daff.gov.au/__data/assets/pdf_file/0010/931609/report-single.pdf [accessed 13 January 2014].

2 The Royal Society. 2009. *Reaping the benefits: Science and the sustainable intensification of global agriculture*, Policy document 11/09, 8. [Online]. Available at: http://royalsociety.org/uploadedFiles/Royal_Society_Content/policy/publications/2009/4294967719.pdf [accessed 13 January 2014].

3 Bragdon, S. 2013. *Small-scale farmers. The missing element in the WIPO-IGC draft articles on genetic resources*. QUNO Quaker United Nations Office, Briefing Paper, 7 [Online]. Available at: http://www.quno.org/sites/default/files/resources/QUNO%20Genetic%20resources%20Farmers%20Food%202013.pdf [accessed 13 January 2014].

small scale farmers 'to create new and relevant varieties, and maintain diversity on-farm, as well as to mix new varieties with traditional varieties' will be crucial to future food security.[4]

Controversies about the introduction of GMOs challenge the mainstream understanding of 'traditional', highlight uncertainty about the characteristics assessed in measuring what is traditional and what is not, affect how equivalence or the lack of it are assessed and force us to confront questions about how we should define legal concerns around modern biotechnology. The assertion that GMOs are only an extension of centuries-old human-directed genetic change in agricultural plants and animals is well known. So too is the response that GMOs are not traditional because they are a product of a more modern form of biotechnology (or GM) that allows a broader array of genetic traits to be transferred between organisms than conventional breeding.[5] Whatever the case, agricultural GMOs and the legal developments accompanying them, particularly the growing role of intellectual property in agriculture, are often criticised on the basis that they threaten tradition and inhibit traditional innovation. At the same time, legal tools are increasingly being used to protect 'tradition'. However, little consideration has been given to the space allowed for traditional innovation in these regimes and what they mean for the adoption of new technologies such as GM.

This chapter first considers what tradition means in the context of agricultural innovation. Developments in the legal protection of tradition relevant to agricultural genetic change are the subject of the second section. The first part of the second section considers developments in the protection of geographical indications (GIs) and the second part considers developments directed at preserving traditional knowledge (TK). The results of that analysis are then used to critically assess opportunities for traditional innovation in agricultural organisms and what such opportunities entail for the adoption of GMOs.

Tradition

Hundreds of years of ongoing genetic change, driven by traditional innovation, illustrates that both the undertaking and products of genetic change are acceptable to the traditional lifestyle of some societies. Loss of the ability to develop new varieties through traditional innovation could itself be considered a loss of tradition, in the same manner as the genotype of agricultural varieties and/or the methods used to produce those varieties can be considered

---

4   Ibid., 2.

5   At a very simplistic level, conventional breeding is limited by the available genetic variability in the target organism and its sexually compatible relatives, whereas GM removes these limits, allowing traits from far removed species to be combined.

'traditional'. Nevertheless, new technology such as GM is sometimes rejected as destroying tradition. Lyria Bennett Moses explains that 'new technology carries with it new possibilities, and these can potentially conflict with existing social, environmental and cultural values.'[6] However, what leads some losses to be called 'cultural losses' is unclear.[7] There is no agreed definition of 'tradition' and care must be taken not to impose one culture's view onto another's when deciding on what is and is not included in that term.[8] The purposes of visualising traditions in some circumstances, such as establishing a right to compensation pursuant to Western legal systems, may also mean that innovation is overlooked to satisfy such systems' demand for a continuing and unchanging culture.[9]

Policymakers are increasingly using legal tools to protect tradition from new technologies and innovations and such responses are of growing international interest.[10] As Tomer Broude asserts, 'if aggregate individual consumer demand cannot independently sustain the cultural widget in the face of non-cultural but otherwise functionally substitutable products, the widget's economic survival requires regulatory protection for its preservation.'[11] A difficulty for policymakers though is that 'it is unclear whether all aspects of a people's culture are equally valuable or by what criteria – to say nothing of by what agency – that shall be determined,'[12] The answer to this will be impacted by the purpose of the decision-making. For example, different criteria may be relevant where the objective is biodiversity protection compared to where the objective is to create a more equitable regime for traditional societies or, again, where it is to enhance trade. In light of these different objectives, it may be that approaches to traditional innovation in one regulatory arena are inappropriate in others.

---

6   Bennett Moses, L. 2007. Recurring dilemmas: The law's race to keep up with technological change. *University of Illinois Journal of Law, Technology and Policy*, 7(2), 239–85, 248.

7   Dominguez, V. 2001. Comment on Kirsch, S. Lost worlds: environmental disaster, 'cultural loss' and the law. *Current Anthropology*, 42(2), 182–3, 183.

8   For a discussion of the difficulties caused by the lack of definitional boundaries regarding traditional knowledge, see Phillips, P. 2014. Traditional Knowledge, in *Socio-Economic Considerations in Biotechnology Regulation*, edited by K. Ludlow, S. Smyth and J. Falck-Zepeda. New York: Springer.

9   Kirsch, S. 2001. Lost worlds: Environmental disaster, 'cultural loss' and the law. *Current Anthropology*, 42(2), 167–78, 177.

10   Ibid., 171.

11   Broude, T. 2005. Taking 'trade and culture' seriously: Geographical indications and cultural protection in WTO law. *University of Pennsylvania Journal of International Economic Law*, 26(4), 623–92, 646.

12   Rosen, L. 2001. Comment on Kirsch, S. lost worlds: environmental disaster, 'cultural loss' and the law. *Current Anthropology* 42(2), 188–9, 189.

Genetic Change and GIs

Policy Objectives

A common justification for GIs is their potential to protect tradition and local culture. For example, Daphne Zografos observes that GIs today are seen 'as a legal and economic tool for the development of rural areas and the preservation of cultural heritage'.[13] Similarly Sarah Bowen notes that GIs help maintain a diversity of local food cultures because 'they protect the unique traditions, knowledge, and environmental resources that are associated with typical products.'[14] Further, GIs have an important part in claims that 'European agriculture is "traditional" while agricultural production in North America, Australia, and other New World countries is "industrial".'[15] The inclusion of GIs within the World Trade Organisation's realm through the *Agreement on Trade-related Aspects of Intellectual Property Rights* (TRIPS) however, makes GIs a legal tool within international trade with objectives compatible with that arena. As discussed below, GI regimes arguably preserve the state of particular cultures and allow less space for traditional innovation than the proposed approaches to TK protection discussed in the following section.

The relevance of GIs for the protection of TK has been explored by several commentators. Zografos, for example, looks at the relevance of GIs for the economies of developing countries that use GIs to protect TK and reward holders of that knowledge.[16] However, she does not consider the impact of GIs on the local or indigenous community's ability to innovate. The risk is that, as Justin Hughes has observed, '[l]ike rights to folklore or traditional knowledge, geographical indications crystallize protection around traditional purveyors/creators without regard to recent originality or creativity.'[17] Similarly, Broude highlights that the notion of 'authentic' food artificially attempts to crystallise or capture culture at a particular moment, when in fact culture is constantly changing and evolving.[18]

---

13   Zografos, D. 2008. *Geographical indications and socio-economic development.* Working Paper 3, Dec, IQsensato, iii. [Online]. Available at http://www.iqsensato.org/?page_id=49 [accessed 14 January 2014].

14   Bowen, S. 2010. Development from within? The potential for geographical indications in the global South. *Journal of World Intellectual Property*, 13(2), 231–52, 234.

15   Hughes, J. 2006–2007. Champagne, feta, and bourbon: The spirited debate about geographical indications. *Hastings Law Journal*, 58(2), 299–386, 339.

16   Zografos, above n 13. See also Dutfield, G. 2013. Geographical indications and agricultural community development: Is the European model appropriate for developing countries, in *The Intellectual Property and Food Project. From Rewarding Innovation and Creation to Feeding the World*, edited by C. Lawson and J. Sanderson. UK: Ashgate Publishing Ltd, 175–200.

17   Hughes, above n 15, 335 (footnote omitted).

18   Broude, above n 11, 18–21.

In contrast, other commentators consider that GIs are able to accommodate innovation. Graham Dutfield has observed that the 'characteristics of the product and manufacturing techniques and conditions will not remain the same indefinitely but will evolve. Such evolution may entail the abandonment of some long-established cultural norms where these hinder the scale-up of production to meet increased market demand. Obviously, securing the integrity of the indication during such change is a vital consideration.'[19] Hughes has also noted examples that show GIs have accommodated genetic change, such as the grafting of grape rootstocks and varietals and the changing constitution of the cattle stock used to produce Parmigiano-Reggiano cheese.[20] In the latter case, the entire cattle herd used to produce cheese was changed over from local cattle to descendants of North American Holsteins and Dutch Friesians. However, as Hughes observes, these examples may suggest a disconnect with the justifications for GIs.[21]

Definition of GIs

As noted above, GIs are part of the international intellectual property regime. TRIPS obliges member countries to protect GIs although this can be done through differing domestic measures. For the purposes of TRIPS, GIs are defined as:

> indications which identify a good as originating in the territory of a Member, or a region or locality in that territory, where a given quality, reputation or other characteristic of the good is essentially attributable to its geographical origin.[22]

Analysis

As with regimes aimed at protecting TK discussed below, equivalence (or rather the lack of it) is crucial to the justification of the exclusion of some products from GI protection. However, how equivalence is to be measured is unclear. As other commentators have explained, the express requirement that a particular quality, reputation or other characteristic of the good be *essentially attributable* to its geographical origin is ambiguous. It is not clear whether and to what extent human production factors may also be part of that quality, reputation or other characteristic.[23] After consideration of the history of the drafting of the TRIPS text, Hughes suggests that human factors of production can be included, although there still must be an essential land/qualities connection.[24] However, as Melissa de

---

19   Dutfield, above n 16, 187.

20   Hughes, above n 15, 360.

21   Ibid.

22   *Agreement on Trade-related Aspects of Intellectual Property Rights*, 33 ILM 1197 (1994), Art 22.1.

23   Hughes, above n 15, 315; Dutfield, above n 16, 180–1.

24   Hughes, ibid.

Zwart notes, how this is to be measured or assessed is not prescribed.[25] Nor, it is asserted here, is it clear what is included within the concept of geographical origin and, in particular, whether that includes a locality's genetic heritage, static or not.

This is important to this chapter's exploration of the boundaries between tradition and innovation for two reasons. First, the contribution to a good's qualities by its geographical origin in contrast to human factors must be measurable if the linking to comparative inputs is to be meaningful. Secondly, it raises the issue of whether any agricultural organism grown in a particular locality can satisfy this requirement regardless of genetic change and the technique used for directing that change. In regards to assessment of the contribution of geographical origin as distinct from other factors, the inputs and outputs measured to decide this and how they are measured will be fundamental. As the science of genetics develops, many attributes of organisms or their products will be attributable to genotype and perhaps the effect of the environment on it. This already is or will become in the future replicable by GM, synthetic biology or other modern biotechnology techniques. A conclusion that genetic makeup must remain static provides no space for traditional or modern innovation through genetic change. But if genotype does not have to remain static to gain GI protection, GMOs or synthetically created organisms could perhaps be substituted for those agricultural organisms presently used. The method used to produce that genotype, whether traditional or modern, seems more obviously to be a human factor rather than a factor of geographical origin.

The obligations imposed on members with respect to GIs may assist in answering these questions. Member's obligations in relation to GIs are described in TRIPS Art 22.2 that provides:

> Members shall provide the legal means for interested parties to prevent:

> the use of any means in the designation or presentation of a good that indicates or suggests that the good in question originates in a geographical area other than the true place of origin in a manner which misleads the public as to the geographical origin of the good;

> any use which constitutes an act of unfair competition within the meaning of Art 10*bis* of the Paris Convention (1967).

Article 22.2(a) is limited to the prevention of deception as to a good's geographical origin and not as to its other characteristics. The same uncertainties raised by this limitation are raised by the definition of GI itself. However, Art 22.2(b) is not so limited. Article 10.3 of the *Paris Convention for the Protection of Industrial*

---

25   de Zwart, M. 2012. Geographical indications: Europe's strange chimera or developing countries' champion?, in *The Law of Reputation and Brands in the Asia Pacific*, edited by A.T. Kenyon, N-L. Loon and M. Richardson. UK: Cambridge University Press, 233–52, 236.

Property (Paris Convention 1967), to which Art 22.2(b) refers, prohibits 'indications or allegations the use of which in the course of trade is liable to mislead the public as to the nature, the manufacturing process, [or] the characteristics' of the goods.[26] Whether this list includes the genetic heritage of an agricultural product will depend upon the domestic measures introduced in response to TRIPS obligations.

Looking at the domestic measures introduced by members so far, some use GIs to protect genotype and others do not. The United States and Australia, for example, do not mandate varietals or control production conditions. GI quality controls are instead principally a matter for market forces.[27] However, other nations have GI regimes that would or could restrict genetic innovation or the use of GM techniques. For example, the French Appellation of Control (AOC) regime extends to the natural and human factors of the geographic environment of the product's source, which is wider than the TRIPS reference to geographical origin.[28] De Zwart notes that the French AOC has 'a rigorous set of controls over the variety of grapes that can be harvested'.[29] Other nations have also adopted GI systems that confine the genetic change that is allowed, even if only at the variety rather than gene level. For example, Dutfield describes a Mexican example where the specification for a particular GI, Tequila, permits only one crop species or variety to be used where others could reasonably be used as well.[30]

In nations such as France, the use of GI regulations to restrict production methods is justified on the basis of the effect of *terroir*. Hughes observes that *terroir* is an:

> input/output idea: some unique inputs (the terroir elements) produce unique outputs in the same way that individual artisans might produce stylistically unique outputs. The particular input is *necessary* for the particular output: no other input produces that output.[31]

However, as with the definition of GI, whether *terroir* is responsible for the product's characteristics depends upon the characteristics measured. According to Hughes, it includes geology and climate, to which advocates add culture, history and human skill. Hughes notes on the input side that '[i]n a trade environment that demands a certain level of science, rationality, and transparency, if terroir

---

26   *Paris Convention for the Protection of Industrial Property* 828 U.N.T.S. 305, Art 10.3.

27   Hughes, above n 15, 333–4.

28   Gervais, D. 2009. Traditional knowledge: Are we closer to the answer(s)? The protectional role of geographical indications. *ILSA J of International and Comparative Law*, 15(2), 551–67, 559.

29   De Zwart, above n 25, 239.

30   Dutfield, above n 16, 188.

31   Hughes, above n 15, 357.

remains "an article of faith" for some but a dubious mystery to many, it cannot be a useful concept for developing further international norms.'[32] Allowing GIs to change across time runs the risk of threatening the uniqueness of the inputs but not allowing change means innovation, traditional or modern, is stifled.

As Bowen has observed, a significant deficiency in the use of GIs in the protection of tradition occurs when governments fail to establish objectives for GI policies beyond using GIs as a trade barrier to protect their country's products from foreign-produced imitations.[33] Further, the lack of decision-making criteria around changes to any particular GI for the institution or organisation responsible for managing GIs or GI policy means the process of norm setting is hidden and can be dominated by interest groups.[34] Finally, in the context of the issues raised by this chapter, the lack of a framework for measuring equivalence makes conclusions reached, even if there are decision-making criteria, difficult to justify. These factors mean that GI regulation can be captured by groups which, intentionally or not, do not allow space for innovation in the form of genetic change by farmers. This is a particular concern where rules from developed countries are transported to developing countries, where innovation through genetic change may be of far greater significance, whether economic or cultural.

Genetic Change and Traditional Knowledge

Commentators have explained that tradition in the context of TK is not determined by the age of the knowledge but by how it is acquired and used. This then means there is space for innovation. Russel Barsh explains that:

> the social process of learning and sharing knowledge, which is unique to each indigenous culture, lies at the very heart of its 'traditionality'. Much of this knowledge is actually quite new, but it has a social meaning and legal character, entirely unlike the knowledge indigenous peoples acquire from settlers and industrialized societies.[35]

Cynthia Ho's explanation is also useful. She recognises that '[TK] is constantly evolving over generations and in response to a changing environment ... communities strongly value the development and preservation of their environment,

---

32   Ibid., 358 (footnotes omitted).

33   Bowen, above n 14, 243.

34   Ibid., 242.

35   Barsh, R. 1999. Indigenous knowledge and biodiversity, in *Cultural and Spiritual Values of Biodiversity*, edited by D. Posey. London: United Nations Environmental Programme and Intermediate Technology Publications, 73–6, 74–5, and cited in Dutfield, G. 2001. TRIPS–related aspects of traditional knowledge. *Case W Res J Intl L* 33(2), 233–75, 242.

including plant and agricultural resources.'[36] As an example of such development Ho notes that '[TK] is considered responsible for helping communities develop diverse crops through knowledge and cultivation of their environment'[37] and 'there is a naturally self-sustaining incentive to protect, promote, and continue to improve such knowledge.'[38]

International responses to TK generally refer to the possibility of development or innovation. For example, the UN *Declaration on the Rights of Indigenous Peoples* provides for the right to TK protection whilst also envisaging the development of it:

> Indigenous peoples have the right to maintain, control, protect and develop their cultural heritage, traditional knowledge and traditional cultural expressions, as well as manifestations of their sciences, technologies and cultures, including human and genetic resources, seeds, medicines, knowledge of the properties of fauna and flora, oral traditions, literatures, designs, sports and traditional games and visual and performing arts. They also have the right to maintain, control, protect and develop their intellectual property over such cultural heritage, traditional knowledge, and traditional cultural expressions.[39]

Important international fora addressing TK protection include the World Intellectual Property Organisation (WIPO) and the UN *Convention on Biological Diversity* (CBD). WIPO is currently negotiating IP rules to protect TK. The CBD is relevant because it obliges members to protect, amongst other things, TK as part of its objectives to conserve and protect biodiversity. Innovation is expressly included in discussions in both of these arenas although there is no definition of that term. Relevant developments in both these fora are considered below.

World Intellectual Property Organisation

The mandate of WIPO's Intergovernmental Committee on Intellectual Property and Genetic Resources, Traditional Knowledge and Folklore (IGC) was renewed at the recent WIPO General Assembly held from 23 September to 2 October 2013. Pursuant to that mandate, the IGC is to continue negotiations with the objective of reaching agreement on a text (or texts) of an international legal instrument(s) to ensure the effective protection of TK. That text is to be submitted to the 2014 General Assembly, where a decision will be made as to the convening of a

---

36   Ho, C. 2006. Biopiracy and beyond: A consideration of socio-cultural conflicts with global patent policies. *University of Michigan Journal of Law Reform* 39(3), 433–542, 446–7 (footnotes omitted).

37   Ibid., 447 (footnotes omitted).

38   Ibid., 448 (footnotes omitted).

39   *United Nations Declaration on the Rights of Indigenous Peoples* A/RES/61/295, Art 31.

Diplomatic Conference. Related texts on genetic resources and traditional cultural expressions are also being developed by the Committee but these, although relevant, lie beyond the scope of this chapter.

TK is to be the focus of the IGC's twenty-seventh meeting in April 2014, where work is to continue on a revised draft text. The text is presently largely unsettled and important basic points, including the definition of TK, have still to be agreed.[40] As it stands, Art 1 attempts to define the protected subject matter, namely TK. Article 3 provides for the scope of protection to be given to TK and Art 6 provides for exceptions and limitations to that protection. The following comments relating to innovation through genetic change can be made.

Policy objectives
The policy objectives of the current draft clearly allow and intend innovation to occur. Two stated objectives for TK protection are particularly relevant: recognition of value and promotion of innovation. In regards to the first noted objective, recognition of value, the current draft says that it should be recognised that '[TK] systems are frameworks of ongoing innovation ... and have equal scientific value as other knowledge systems.'[41] Space for innovation is therefore clearly intended. The text around the objective of promoting innovation is far more contentious, with a number of alternative texts still being considered. One alternative suggests that the objective is to 'encourage and reward ... tradition-based creativity and innovation'.[42] This would seem to refer only to innovation by indigenous and local communities. An alternative version though provides that TK protection is intended to 'promote innovation, creativity and the progress of science, and promote the transfer of technology on mutually agreed terms'.[43] This alternative may aim to promote innovation by the indigenous and local communities concerned, those outside such communities or both. A third version more clearly includes an encouragement of innovation by both holders and users of TK by providing that 'the protection of TK should contribute toward the promotion of innovation and to the transfer and dissemination of knowledge to the mutual advantage of the holders and users of TK and in a manner conducive to social and economic welfare and to a balance of rights and obligations.'[44]

Neither the recognition of the equal scientific value of traditional and other innovation or the promotion of traditional innovation justifies or requires the rejection of modern technology. Indeed, if the wording of the final possibility for

---

40   World Intellectual Property Organisation, *The Protection of Traditional Knowledge: Draft Articles Rev 2* (14 May 2013), WIPO/GRTKF/IC/24/FACILITATORS DOCUMENT REV. 2. [Online]. Available at http://www.wipo.int/meetings/en/doc_details. jsp?doc_id=238182 [accessed 14 January 2014].

41   Ibid., Policy Objectives, (i), (p 3.

42   Ibid., Policy Objectives, (x), 4.

43   Ibid., Policy Objectives, (x, Alternative), 4.

44   Ibid., Policy Objectives, (xx, Alternative), 6.

the promotion of innovation is adopted, that promotion is to be undertaken 'in a manner conducive to social and economic welfare'.[45] There is no guidance so far on how welfare would be determined but it cannot be assumed that welfare is always best served by rejecting modern technologies. Indeed the first of the two objectives for the creation of WIPO itself is the promotion of intellectual property[46] and, whilst the purpose of that protection is not stated, WIPO's Strategic Goals include facilitating the use of IP for development.[47]

Definition of TK
Draft Art 1.1 provides that TK:

> [refers to]/[includes]/[means] know-how, skills, innovations, practices, teachings and learnings of [indigenous [peoples] and [local communities]]/[or a state or states] that are dynamic and evolving, and that are intergenerational/and that are passed on from generation to generation, and which may subsist in codified, oral or other forms.

> [Traditional knowledge may be associated, in particular, with fields such as agricultural, environmental, healthcare and indigenous and traditional medical knowledge, biodiversity, traditional lifestyles and natural resources and genetic resources, and know-how of traditional architecture and construction technologies.] (footnote omitted).[48]

Article 1.2 goes on to define TK associated with genetic resources:

> [Traditional knowledge associated with genetic resources means [substantive] knowledge of the [properties], and uses of genetic resources and their derivatives held by indigenous [peoples] and local communities [and which directly leads to a claimed invention].][49]

These provisions make it clear that TK can include innovations, including innovations associated with genetic resources and their use in agriculture. However, the current version of Art 1.3 regarding criteria for eligibility for protection may limit the space for innovation if a minimum time threshold is adopted. Draft Art 1.3 provides:

---

45   Ibid., Policy Objectives, (xx, Alternative), 6.
46   *Convention Establishing the World Intellectual Property Organisation* 828 U.N.T.S. 3, Art 3.
47   *WIPO Strategic Goals.* [Online]. Available at: http://www.wipo.int/about-wipo/en/goals.html [accessed 13 February 2014].
48   WIPO/GRTKF/IC/24/FACILITATORS DOCUMENT REV. 2, above n 40, Art 1.1, 10.
49   Ibid., Art 1.2, 10.

Protection extends [only] to traditional knowledge that is [distinctively] associated/ linked with the cultural, [and] social identity, [and] or cultural heritage of beneficiaries as defined in Art 2, that is generated, maintained, shared/ transmitted in collective context, that is intergenerational/that is passed on from generation to generation [and has been used for a term as may be determined by each [Member State]/[Contracting Party] but of not less than [fifty years]] [recognizing the [cultural] diversity of the beneficiaries] recognizing that there is cultural diversity amongst beneficiaries (footnotes omitted).[50]

Depending upon the final wording, innovations may therefore need to be of a certain age, at present at least 50 years old, to be eligible for protection. Presumably until the relevant age is reached, any innovation would not be protected as TK although it may be protected under other IP laws. The innovation must also be linked in some way with the culture or cultural heritage of the indigenous and local communities.

Analysis

If a particular innovation falls within the protected subject matter, the scope of protection is described in Art 3.1 that is, in part:

[Member States]/[Contracting Parties]/[This instrument] [should]/[shall] confer(s) the following [exclusive] [collective] rights on the beneficiaries, as defined in Art 2:

(a) to maintain, control, [protect][51] and develop their [protected] [secret] traditional knowledge ...[52]

Article 3.1 is directed at what rights the holder of TK has, which include developing that TK, rather than when the developments themselves are also to be treated as TK.

A draft exclusion from protection, provided in Art 6.9, is particularly relevant to debates around GMOs, whether they are 'traditional' and their effect on traditional innovation. Article 6.9 crystallises, without answering, many of the uncertainties around the distinction between tradition and innovation. Draft Art 6.9 provides:

[[There shall be no right to [exclude others] from using knowledge that:]/[The provisions of Art 3 shall not apply to any use of knowledge that:]

---

50   Ibid., Art 1.3, 10.

51   As Bragdon observes in the context of the ability of small-scale farmers to create new genetic resources on-farm, it is not clear what 'protecting' genetic resources means. Bragdon, above n 3, 5.

52   WIPO/GRTKF/IC/24/FACILITATORS DOCUMENT REV. 2, above n 40, Art 3.1, 12.

(a) has been independently created [outside the beneficiaries' community];

(b) [legally] derived from sources other than the beneficiary; or

(c) is known [through lawful means] outside of the beneficiaries' community.][53]

The effect of Art 6.9 may be that where a GMO (or more correctly, the knowledge around an organism's genetic makeup) is equivalent to a traditional innovation the latter will not be protected if the GMO (or knowledge about it) has been independently created, derived from other sources or is otherwise known. However, how equivalence is to be measured for these purposes is not provided for. There are different ways to 'know' the same thing. Equivalence in the context of genetic change to an agricultural organism could be assessed on the basis of phenotype, measurable change in DNA or proteins present in the organism, the technique used to bring about the genetic change, the place of the organism or the food it produces in the culture of the group concerned or something else altogether.

The United Kingdom's House of Lords has considered the meaning of 'knowing' in everyday life in modern society, as compared with a traditional culture, as part of a decision on novelty for the purposes of patent law.[54] Lord Hoffmann identified the difficulty in comparing equivalence of knowledge possessed by the two groups by noting that:

> [t]here is an infinite variety of descriptions under which the same thing may be ·
> known. Things may be described according to what they look like, how they are
> made, what they do and in many other ways. Under what description must it be
> known in order to justify the statement that one knows that it exists?[55]

Lord Hoffmann goes on to illustrate the problem by comparing the knowledge of Amazonian Indians with that of modern society regarding quinine. He notes that both know about quinine's usefulness in treating malarial fevers: the Indians by virtue of their animistic knowledge that a certain bark can treat such fevers and modern science through the isolation, chemical characterisation and artificial synthesis of the active ingredient in that bark, quinine.[56] The same issue arises in regards to draft Art 6.9 – what does it mean to know something or independently create it? When will a genetic modification, synthetically created DNA sequence or organism (or knowledge about these) be equivalent to a traditional innovation produced in a conventional TK setting?

---

53   Ibid., Art 6.9, 21.

54   *Merrell Dow Pharmaceuticals Inc v. HN Norton & Co Ltd* [1995] UKHL 14 [39], (1996) 19(1) IPD 19004.

55   Ibid., [36].

56   Ibid., [37] cited in Dutfield, above n 35, 257.

Convention on Biological Diversity

Article 8(j) of the CBD sets an international agenda on TK.[57] This provision expressly refers to innovations although, once again, there is no definition of that term. It requires the CBD's 193 Member States, as far as possible and as appropriate, to preserve TK by providing that each member:

> Subject to its national legislation, respect, preserve and maintain knowledge, innovations and practices of indigenous and local communities embodying traditional lifestyles relevant for the conservation and sustainable use of biological diversity and promote their wider application with the approval and involvement of the holders of such knowledge, innovations and practices and encourage the equitable sharing of the benefits arising from the utilization of such knowledge, innovations and practices.[58]

Policy objectives

The objectives of the CBD are defined as 'the conservation of biological diversity, the sustainable use of its components and the fair and equitable sharing of the benefits arising out of the utilization of genetic resources'.[59] These objectives limit the scope of the CBD itself whilst its provisions provide members with wide-ranging discretion as to implementation. Nevertheless, it is still contested how far, if at all, beyond association with genetic resources the protection of TK is to go. Of particular importance is the implementation of the CBD's *Nagoya Protocol on Access to Genetic Resources and the Fair and Equitable Sharing of Benefits Arising from their Utilization to the Convention on Biological Diversity* (Nagoya Protocol) which addresses access and benefit sharing.[60] TK in the context of the Nagoya Protocol is limited to TK associated with genetic resources and this means the priority of other TK issues in future work on Art 8(j) is contested.[61]

The strategy for meeting the CBD's objectives is set out in the *Strategic Plan For Biodiversity For The Period 2011–2020. 'Living in Harmony With Nature'*.[62] This Plan includes 20 headline targets (the Aichi Biodiversity Targets), organised

---

57   Importantly, the USA is not a member of the CBD.

58   *Convention on Biological Diversity* 1760 U.N.T.S. 79, Art 8(j).

59   *Convention on Biological Diversity* 1760 U.N.T.S. 79, Art 1.

60   Conference of the Parties to the Convention on Biological Diversity. 2011. *Report of the Tenth Meeting of the Conference of the Parties to the Convention on Biological Diversity*, UNEP/CBD/COP/10/27. Montreal: CBD Secretariat, [103] and Decision X/1, 89–109.

61   IISD Reporting Services, *Summary of the 8th Meeting of the Working Group on Article 8(j) and 17th meeting of the Subsidiary Body on Scientific, Technical and Technological Advice of the Convention on Biological Diversity* 7–18 October 2013. [Online]. Available at http://www.iisd.ca/vol09/enb09611e.html [accessed 22 October 2013].

62   UNEP/CBD/COP/10/27, above n 58, [14] and Decision X/2 Annex, 116.

under five strategic goals. Targets 18 and 19 are particularly relevant to this chapter, providing:

> *Target 18:* By 2020, the traditional knowledge, innovations and practices of indigenous and local communities relevant for the conservation and sustainable use of biodiversity, and their customary use of biological resources, are respected, subject to national legislation and relevant international obligations, and fully integrated and reflected in the implementation of the Convention with the full and effective participation of indigenous and local communities, at all relevant levels.

> *Target 19:* By 2020, knowledge, the science base and technologies relating to biodiversity, its values, functioning, status and trends, and the consequences of its loss, are improved, widely shared and transferred, and applied.[63]

The reconciliation of any tensions between the aspirations in Target 18 to respect TK or traditional innovations with the aspirations in Target 19 of widely sharing and transferring and applying science and technologies relating to biodiversity is not dealt with.

Definition of TK

A Working Group (WG) on Art 8(j) was first established at the fourth meeting of the Conference of the Parties to the Convention on Biological Diversity (May 1998) and its Programme of Work was set in 2000.[64] That Programme lists a number of tasks, some of which have been completed or superseded[65] and others postponed.[66] Amongst the WG's achievements has been the development of the *Akwé: Kon Voluntary Guidelines for the Conduct of Cultural, Environmental and Social Impact Assessments.*[67] Paragraph 6(h) of these guidelines defines TK more narrowly than the definition used by WIPO's IGC, in line with the CBD's objectives. It provides:

> [TK] refers to the [TK], innovations and practices of indigenous and local communities embodying traditional lifestyles relevant for the conservation and sustainable use of biological diversity.[68]

---

63   Ibid., Decision X/2 Annex, 120.

64   Conference of the Parties to the Convention on Biological Diversity. 2000. *Report of the Fifth Meeting of the Conference of the Parties to the Convention on Biological Diversity*, UNEP/CBD/COP/5/23. Montreal: CBD Secretariat, [234] and Decision V/16, 139.

65   Tasks 3, 5, 8, 9 and 16: UNEP/CBD/COP/10/27, above n 60, [421] and Decision X/43 ([5] and [7], 320–21).

66   Tasks 11, 6, 13, 14 and 17. Ibid.

67   Conference of the Parties to the Convention on Biological Diversity. 2004. *Report of the Seventh Meeting of the Conference of the Parties to the Convention on Biological Diversity*, UNEP/CBD/COP/7/21. Montreal: CBD Secretariat, [292] and Decision VII/16F, 261.

68   Ibid., Decision VII/16F, 265.

Further understanding of the place of genetic change in TK for the purposes of the CBD process may be gained from a consideration of projects undertaken under its auspices. The International Institute for Environment and Development (IIED), for example, contributes to the CBD on policy.[69] The IIED is currently undertaking a project with researchers from India, Kenya, China and Peru on smallholder innovation for resilience (SIFOR) through biocultural innovation.[70] The project 'aims to generate thriving smallholder innovation based on traditional knowledge and biocultural heritage (BCH)'.[71] BCH Innovation (BCHI) has been given the following working definition:

> new knowledge, resources, skills and practices, or new combinations of these, which serve to: (a) strengthen and sustain agro-biodiversity, particularly local seed systems, livelihoods and material and spiritual well-being of communities; (b) adapt to and mitigate risks due to global impacts, especially those of climate change. They are practical, sustainable, and are locally and globally relevant.

> BCHIs have their basis in a people's or community's BCH but may incorporate external elements. They integrate daily practices with traditional knowledge, spiritual values and customary norms. As such, they are dynamic, continuous, open, adaptive, and gender-sensitive, integrating the creativity of people and nature.[72]

Amongst other common attributes described as being attributes of a BCHI is that any such development uses a larger proportion of TK than external knowledge and is new to the local area, but not necessarily globally unique. The Inca 'Quipu' system is suggested as an example of a BCHI.[73] This traditional information management system uses knots on string to store information, but is now being used to plot the chromosomes of potatoes and 'as a result, complex scientific information is now accessible to community members in a culturally appropriate format, and modern systems of information dissemination such as the Internet and Web technologies are being used for sustaining TK.'[74] Under the heading of innovations relating to agro-biodiversity and ecosystems, livelihoods and social capital, a Chinese example is described in a workshop report as being the combining of TK and

---

69   IIED. *Who we are.* [Online]. Available at http://www.iied.org/about-us [accessed 10 October 2013].

70   IIED. 2013. *Smallholder Innovation for Resilience: Strengthening biocultural innovation systems for food security in the face of climate change.* Workshop Report 29 April–4 May, Peru.

71   Ibid., 2.

72   Ibid., 14 (footnotes omitted).

73   Ibid., 3.

74   Ibid., 4.

science to develop new maize and rice varieties.[75] This is listed separately from a second example of 'using traditional methods to improve varieties'.[76] It is unclear whether the separation of these two examples is because the researchers involved consider that the use of science is different to improvements made only with traditional methods but in either case the use of science does not prevent the development from being a BCHI. However, the nature of the science and the proportion of its input to the results are not described.

Analysis

The most recent WG meeting (WG8) on 7–11 October 2013 considered the remaining Programme tasks.[77] In the Secretariat's view task 12, an umbrella task,[78] requires the development of guidelines for the development of legislation or other mechanisms, as appropriate, to implement Art 8(j) and its related provisions (which could include *sui generis* systems). There is also a requirement for definitions of relevant key terms and concepts in Art 8(j) and related provisions at international, regional and national levels, that recognise, safeguard and fully guarantee the rights of indigenous and local communities over their traditional knowledge, innovations and practices, within the context of the Convention. WG8 recommended, among other matters, that these tasks be advanced whilst avoiding inconsistencies with the Nagoya Protocol or duplication or overlap with work undertaken in other fora and taking into account relevant developments, including those under the Nagoya Protocol and WIPO's IGC. The language of the WG in making these recommendations does not expressly deal with the development or meaning of traditional innovations but rather with ensuring that indigenous and local communities' rights over innovations be recognised, safeguarded and fully guaranteed.

At WG8 it was agreed that work was, as far as possible, to be carried out in collaboration with other relevant organisations, including WIPO, which had reported on its progress in addressing TK to the WG meeting.[79] This may explain the property centric language used by the WG, focusing on rights over innovation rather than the right to innovate. Similarly, the *Akwé: Kon Voluntary Guidelines* concern the protection of tradition and traditional innovation from 'outside' effects. They do not address how to undertake the assessment of an adverse effect or what makes an innovation traditional. For example, in relation to environmental impact assessments Guideline 36 notes that the direct impacts of a development proposal on local biodiversity should be assessed at the 'ecosystem, species and genetic levels' and Guideline 39 requires 'traditional systems and means of production' to be taken into account, which allow space for traditional innovation. However,

---

75   Ibid., 10.
76   Ibid.
77   Focusing on tasks 1, 2, 4, 7, 10 and 12.
78   IISD Reporting Services, above n 61.
79   UNEP/CBD/COP/5/23, above n 64, Decision V/16, 139–46.

there is no guidance on how assessment is to be done or how equivalence or otherwise is to be measured. Similarly, the guidelines regarding cultural impact assessments require consideration of matters recognising that innovation occurs but do not explain how that assessment is to occur or what is required to achieve the objectives. Guideline 27 for instance, requires consideration of '(a) possible impacts on continued customary use of biological resources; (b) possible impacts on the respect, preservation, protection and maintenance of [TK], innovations and practices'.[80]

Work on assessment of the socio-economic considerations, including TK, arising from the impact of GMOs is being undertaken within a second forum of the CBD arena. An Ad Hoc Technical Expert Group (AHTEG) on Socio-Economic Considerations as addressed by Art 26 of the *Cartagena Protocol on Biosafety* (CPB) to the CBD has been established.[81] This group was to meet for the first time in February 2014 and aimed to 'develop conceptual clarity' on these considerations. The issue of assessment and measuring of equivalence in this forum will, it can be expected, reflect the group's setting within the biosafety field where genetic change is seen as a possible risk but it is also recognised that benefits as well as harm can arise from change. Care will be needed in developing any recommendations and consideration given to how these will interplay with the broader CBD developments on TK and, in turn, how the CBD developments interact with other legal regimes aimed at protecting tradition.

Conclusions

Agriculture is the dynamic management of 'the interaction between crop genotypes or livestock breeds and their immediate agro environment (physical and biological)'.[82] Innovation is an essential part of that dynamic process. There is a risk that existing GI regimes and proposed legal responses to TK will stifle innovation, leading to farmers and communities having to continue using outdated technology and genotypes. Whilst modern agricultural biotechnology is sometimes challenged as threatening tradition, the stifling of traditional innovation may itself be a loss of tradition.

The legal responses to tradition considered in this chapter all recognise that traditional innovation has always occurred, with the proposed legal responses

---

80   See also Guidelines 24, 28 and 29: UNEP/CBD/COP/7/21, above n 67, Decision VII/16F, 269.

81   Conference of the Parties to the Convention on Biological Diversity Serving as the Meeting of the Parties to the Cartagena Protocol on Biosafety. 2012. *Report of the Sixth Meeting of the Conference of the Parties to the Convention on Biological Diversity Serving as the Meeting of the Parties to the Cartagena Protocol on Biosafety*, UNEP/CBD/BS/COP-MOP/6/18. Montreal: CBD Secretariat, [186] and Decision BS-VI/13, 93.

82   The Royal Society, above n 2, 7.

being developed by WIPO and the CBD members expressly referring to innovation. The GI regime does not expressly address innovation. Nevertheless, most commentators consider that the GI regime allows for innovation and that care must be taken when writing or amending GI specifications to allow for this.

The lack of clarity around the objectives of the considered regimes, however, makes it difficult to assess the actual space intended to be left for traditional innovation. Importantly for this chapter, it also makes it difficult to predict what these regimes mean for the adoption of new technologies such as GM. As noted above, all three regimes aim to protect tradition and provide space for innovation, although the reasons for that protection are not clear. Some guidance can be drawn from the overarching objectives of the regimes – the overall objective for the WIPO regime being promotion of intellectual property, for the CBD members conservation and sustainable use of biodiversity and benefit sharing and, for the GIs regime, part of the laws are about international trade.

In regards to GIs, many countries with GIs that restrict innovation in production methods, such as the EU, also have comparatively restrictive regimes on GMO adoption. Whilst this may be an attempt to maintain sovereignty in international trade or conserve biodiversity, there is lack of clarity around how to measure the significance of geographical origin and therefore predict what innovation is acceptable.

The WIPO regime also intends to promote, and protect, traditional innovation. However, depending upon the final text adopted, protection (and therefore promotion) may be removed from traditional innovations if an 'equivalent' is independently created through modern biotechnology. Discussions so far on the draft text make it likely that laboratory created knowledge can be equivalent to TK created knowledge. However, as with GIs, how equivalence is to be measured is unclear. To expand on Lord Hoffmann's discussions, if an indigenous group knows that a particular variety has certain attributes and a laboratory can explain why the variety has those attributes, and can perhaps even replicate those attributes using biotechnology, do they know the same thing? These uncertainties seem to be the very crux of debates over whether GMOs are traditional or, at least, only an extension of the traditional. It is very likely different groups will have different answers to that question depending upon the reason the question is being asked. Lack of clarity in the broader goals of TK protection in the WIPO arena makes it difficult to predict what characteristics will be measured in deciding this.

The CBD regime has a narrow definition of TK, reflecting the CBD objectives, although like the WIPO developments, its focus is the protection of traditional innovation. The restriction on the WG's work to avoid duplication with work in other fora means that decisions regarding assessments of both equivalence and ramifications of the introduction of modern biotechnology may be left to the AHTEG on Art 26 of the Cartagena Protocol rather than the WG on CBD Art 8(j).

Better consideration of how the law should allow for traditional innovation should improve understanding of the legal concerns arising regarding GMOs. In regard to claims that GM threatens tradition it may be that the resulting artefact of human directed genetic change, namely the plant or animal or its particular use, is not the relevant comparator. Rather, the cultural value of an agricultural plant or animal may lie, as Broude observes, in 'the method of production and the lifestyle that both supports it and is supported by it'.[83] Such an approach justifies the rejection of assertions made by some GM advocates that because the same phenotypes, if not genotypes, can be produced using either traditional innovation or modern biotechnology, GMOs and the food they produce should not be seen as challenges to tradition. However, if the 'traditional' technology is the thing to be protected, care will need to be taken in creating legal protection for it to ensure that there is space for traditional innovation. Otherwise the legal tools aimed at protecting tradition may themselves stifle genetic change, modern and traditional.

---

83   Broude, above n 11, 10–11.

Chapter 6

Myriad Genetics and the Remaining Uncertainty for Biotechnology Inventions

Dianne Nicol

Introduction

There have been ongoing academic and policy debates around the question of what constitutes patent eligible subject matter in the field of biotechnology since Professor Ananda Chakrabarty and other researchers filed the first patents for genetically engineered microrganisms. In Chakrabarty's case, the invention at issue was a bacterium that had been engineered to degrade crude oil through the introduction of four types of plasmids that carried relevant genes.[1] His application was initially rejected by the United States Patent and Trademark Office (USPTO).[2] In 1980, some six years after the key scientific papers describing Chakrabarty's bacterium had been published the Supreme Court of the United States (SCOTUS) was given the opportunity to rule on its patent eligibility. In *Diamond v. Chakrabarty*,[3] the Court held that there was nothing intrinsically patent ineligible about the microorganism. While the Court recognised that 'laws of nature, physical phenomena, and abstract ideas' lacked eligibility,[4] the majority held that living organisms could be patented if they exhibited 'markedly different characteristics from any found in nature'.[5]

From 1980 up to the end of the first decade of the twenty-first century, US law on patent eligible subject matter relating to products and processes from the natural world essentially stood still. The only concrete exceptions were a decision of the Board of Patent Appeals and Interference confirming the patent eligibility of plants in 1985 and a rather more significant SCOTUS decision affirming this in 2001.[6]

---

1   Chakrabarty, A. and Friello, D. 1974. Dissociation and interaction of individual components of a degradative plasmid aggregate in Pseudomonas. *Proceedings of the National Academy of Science USA*, 71(9), 3410–14.

2   As reported in the decision of the Supreme Court of the United States, *Diamond v. Chakrabarty*, 447 U.S. 303, 303(1980).

3   Ibid.

4   Ibid., 309.

5   Ibid., 310.

6   In *Ex parte Hibberd*, 227 USPQ 443 (1985) the Board of Patent Appeals and Interference held that plants could be patent eligible, despite also being eligible for plant

This situation changed dramatically in 2010, when a challenge was made to the validity of patents claiming products and processes relating to the BRCA genes, two human genes which had been shown to be linked with hereditary forms of breast cancer. In many respects, these patents were an ideal target for public interest litigation. Breast cancer is an insidious and often fatal disease, although with early diagnosis or pre-symptomatic assessment of risk prophylactic measures can be taken.[7] Myriad Genetics, the holder of the relevant patents had a reputation for aggressive patent enforcement, increasing the cost of testing for relevant mutations and reputedly leading to denial of access to testing for some patients.[8] Lack of availability of second opinion testing was seen as particularly problematic.[9] Thus, the scene was set for a challenge to the patent eligibility of the BRCA genes and methods for utilising them in diagnostic testing.[10] The case was initiated by the American Civil Liberties Union, in conjunction with the Public Patent Foundation on behalf of the Association of Molecular Pathology and others.

The first part of this chapter discusses the ensuing litigation, culminating in the SCOTUS decision in *AMP v. Myriad Genetics* (AMP SCOTUS).[11] The AMP SCOTUS decision warrants consideration in and of itself in a book about agricultural biotechnology, because of the potential it has to impact on the eligibility of patents for genes in this sector (and potentially any other subject

---

variety rights protection under the *Plant Patent Act* and the *Plant Variety Protection Act*. In *JEM Ag Supply v. Pioneer Hi-Bred International, Inc,* 534 U.S. 124 (2001) the Supreme Court confirmed that new plant varieties could satisfy the patentable subject matter requirement.

7   One of the most prominent examples of how these prophylactic measures could be utilised was the double mastectomy undertaken by Angelina Jolie following a positive BRCA test. See, for example, Borzekowski, D., Guan, Y., Smith, K., Erby, L. and Roter, D. 2013. The Angelina effect: Immediate reach, grasp, and impact of going public. *Genetics in Medicine*, doi:10.1038/gim.2013.181.

8   US Secretary's Advisory Committee on Genetics, Health and Society (SACGHS). 2010. *Gene Patents and Licensing Practices and their Impact on Patient Access to Genetic Tests*, 23–4. [Online]. Available at: http://oba.od.nih.gov/SACGHS/sacghsdocuments.html#GHSDOC011 [accessed 6 April 2014]. See also Gold, R. and Carbone, J. 2010. Myriad Genetics: In the eye of the policy storm. *Genetics in Medicine*, 12, S39–S70; Baldwin, L. and Cook-Deegan, R. 2013. Constructing narratives of heroism and villainy: Case study of Myriad's BRACAnalysis® compared to Genentech's Herceptin®. *Genome Medicine*, 5 (8). [Online]. Available at: http://genomemedicine.com/content/5/1/8 [accessed 15 May 2014]; Rai, A. 2014. Diagnostic patents at the Supreme Court. *Marquette Intellectual Property Law Review*, 18(1) 1–9, 3.

9   US Secretary's Advisory Committee on Genetics, Health and Society (SACGHS), ibid., 44.

10   For a list of the patents that were the subject of the legal challenge see Lawson, C. 2014. Patenting DNA sequences after the Myriad decision: New frontiers or just more of the same? *Biotechnology Law Report*, 33(1), 3–16, 5.

11   *Association for Molecular Pathology v. Myriad Genetics, Inc.*, S. Ct. 2107 (2013).

matter derived from the natural world). The way that the decision has since been interpreted by both the judiciary and the USPTO has caused some concern that it may have a more profound impact than originally anticipated, in terms of the broad range of subject matter that could now be patent ineligible. For this reason, the second part of this chapter explores these more recent events in some detail. While this analysis is current at the time of writing, in 2014, it should be read with some caution, given the rapid pace of change.

A Potted History

The lack of a significant body of case law following the Chakrabarty decision meant that the USPTO was left to give its own interpretation of the SCOTUS ruling in that case, without further guidance from the courts. Genetically modified higher organisms, genes, proteins, lipids and other products isolated from the natural world, synthetic processes for their creation in the laboratory and methods for utilising them were all considered by the USPTO to cross the patent eligible subject matter threshold.[12] The USPTO liberally applied the instructions from the court in Chakrabarty that 'anything under the sun made by man' was patent eligible.[13] The task of distinguishing the patentable from the non-patentable was largely left to be resolved through other patent and disclosure requirements. Inventive step and utility, in particular, were relied on to mark out the boundaries around patentable subject matter.[14]

Concerns about the ethics of patenting living organisms started to be raised in the 1980s.[15] However, consideration of the ethics of patenting was alien to many patent examiners and few patent statutes listed this specifically as a matter for consideration. Although the European Patent Convention has an 'ordre public/ morality' exclusion,[16] there has been limited success in utilising this provision

---

12   A number of studies have analysed patent activity. In plant biotechnology, for example, see Graff, G., Rausser, G. and Small, A. 2003. Agricultural biotechnology's complementary intellectual assets. *Review of Economics and Statistics*, 85(2), 349–63; Boettinger, S., Graff, G., Pardey, P., Van Dusen, E. and Wright, B. 2004. Intellectual property rights for plant biotechnology: International aspects, in *Handbook of plant biotechnology* edited by P. Christou and H. Klee. Chichester: John Wiley & Sons Ltd, 1089; Chan, P. 2006. International patent behavior of nine major agricultural biotechnology firms. *AgBioForum*, 9(1), 59–68.

13   *Diamond v. Chakrabarty*, 447 U.S. 303, 309 (1980).

14   On inventive step see *In Re Kubin*, 561 F.3d1351 (Federal Circuit, 2009); on utility see *In Re Fisher*, 421 F.3d 1365 (Federal Circuit, 2005).

15   For example: Dresser, R. 1988. Ethical and legal issues in patenting new animal life. *Jurimetrics Journal*, 28(4), 399–435; Hoffmaster, B. 1988. The ethics of patenting higher life forms. *Intellectual Property Journal*, 4, 1–24.

16   Found in Article 53(a) of the *Convention on the Grant of European Patents* (entered into force 1973) (European Patent Convention – EPC).

to exclude genetic technologies.[17] Arguments for the exclusion of genetic technologies based on morality grounds have gained little traction in recent years and it is posited here that they are unlikely to do so in the future.

It was not until the 1990s that the debate shifted focus, to question the propensity of broad patents over foundational genetic technologies to block off whole areas of research and of thickets of patents to slow the pace of innovation.[18] This was safer ground for debate as it went to the heart of the rationale for allowing patents – to encourage innovation. Yet despite the seemingly flawless logic of these theoretical propositions, blanket bans on patents for genetic technologies were not widely supported in law reform and policy debates, because they were seen as too blunt a solution.[19] On the one hand, freeing up the field by excluding patents for products and processes from the natural world could spur innovation by creating a more open research commons. On the other hand, removal of the patent incentive could deter innovation in this newly emerging field. Even more problematically, it could encourage a culture of secrecy, rather than sharing, creating further barriers to an open research commons.[20]

---

17   The most notable case where the provision has been used successfully is in relation to methods for deriving stem cells from human embryos. See European Court of Justice, Judgement of the Court (Grand Chamber) of 18 October 2011 (reference for a preliminary ruling from the Bundesgerichtshof – Germany) – *Oliver Brüstle v. Greenpeace e.V.* (Case C-34/10) – interpreting Article 6 of the European Parliament and Council, *Directive 98/44/ EC of the European Parliament and of the Council and on the Legal Protection of Biotechnological Inventions* [1998] OJ L 213/13 (the Biotechnology Directive), which was drafted to assist in the interpretation of the EPC in respect of biotechnology inventions. Article 6 assists in the interpretation of Article 53(a) in the EPC. In contrast, in cases prior to the entry into force of the Biotechnology Directive, Article 53(a) of the EPC was not successfully utilised to invalidate patents in relation to: genetically engineered animals, *Oncomouse* T19/90 [1990] *Official Journal of the European Patent Office* 476; genetically engineered plants, *Plant Genetic Systems T356/93* [1995] *Official Journal of the European Patent Office* 545; and human genes, *Relaxin* [1995] *Official Journal of the European Patent Office* 388.

18   On blocking patents see the review by Walsh, J., Arora, A. and Cohen, W. 2003. Effects of research tool patenting and licensing on biomedical innovation, in *Patents in the Knowledge-Based Economy* edited by W. Cohen and S. Merrill. Washington D.C.: National Academies Press, particularly 296–7; on patent thickets and the so-called 'anticommons' that could result see: Heller, M. and Eisenberg, R. 1998. Can patents deter innovation? The anticommons in biomedical research. *Science*, 280(5364), 698–701; Shapiro, C. 2001. Navigating the patent thicket: Cross licenses, patent pools, and standard-setting' in *Innovation Policy and the Economy* edited by A. Jaffe, J. Lerner and S. Stern. Boston: MIT Press, 119.

19   Reviewed in Nicol, D. 2011. Implications of DNA patenting: Reviewing the evidence. *Journal of Law, Information and Science*, 21(1), 7–36.

20   On this point see Cook-Deegan, R., Conley, J., Evans, J. and Vorhaus, D. 2013. The next controversy in genetic testing: Clinical data as trade secrets? *European Journal of Human Genetics*, 21(6), 585–8.

These were and continue to be complex questions. Evidence that might assist in answering them is equivocal at best.[21] With this level of uncertainty, it is hardly surprising that the biotechnology industry has been somewhat reluctant to institute challenges to the patent eligibility of products and processes of nature in the courts.[22] There is a real risk that such challenges could establish precedents that would be damaging to the commercial interests of the industry as a whole. The sequelae to recent judicial decisions, discussed below, illustrate this point. Added to this, there is not a strong tradition of public interest patent litigation in the US, unlike Europe.[23] Taking all these factors into account, it seemed until recently that there was only a remote prospect that a challenge would ever be made to the patent eligibility of genetic technologies based on the patentable subject matter requirement. Lodgement of the complaint by AMP and others challenging the patent eligibility of the BRCA genes created the first opportunity for a judicial ruling on this contentious issue in the US.

The Legal Landscape: The Subject Matter Requirement in International and Domestic Law

Before examining how the courts have interpreted the subject matter requirement in respect of genetic technologies, some brief mention needs to be made of the wording of this requirement in patent legislation. Internationally, countries that are members of the World Trade Organisation are required to be compliant with the *Agreement on Trade-related Aspects of Intellectual Property Rights* (TRIPS).[24] Article 27.1 requires that members make patents available for all inventions in all fields of technology,[25] but provides no further guidance on what constitutes an invention. Articles 27.2 and 27.3 provide three specific exclusions from patenting that member countries are allowed to include in their patent laws. In summary, these provisions cover:

---

21 Caulfield, T., Cook-Deegan, R., Keiff, S. and Walsh, J. 2006. Evidence and anecdotes: An analysis of human gene patenting controversies. *Nature Biotechnology*, 24(9), 1091.

22 Perhaps the closest that the courts came to dealing with this question was in the case of *Amgen Inc v. Chugai Pharmaceutical Company*, 13 USPQ2d 737 (Federal Circuit, Massachusetts District 1989). Although subject matter was not raised as a ground for invalidity, the Federal Circuit intimated that purification and isolation was enough to satisfy the *Chakrabarty* requirement: see Conley, J. 2009. Gene patents and the products of nature doctrine. *Chicago Kent Law Review*, 84(1), 109–32, 116.

23 See, for example, the case law listed above, n17.

24 *Marrakesh Agreement Establishing the World Trade Organisation* (1994) 1867 UNTS 3, Annex IC (*Agreement on Trade-related Aspects of Intellectual Property Rights*) (TRIPS).

25 TRIPS, Article 27.1 adds the proviso that such inventions must be new, involve an inventive step and be capable of industrial application.

- inventions whose commercial exploitation would be contrary to ordre public/morality;[26]
- methods of medical treatment – diagnosis, treatment and surgery on the human or animal body; and
- plants and animals and biological processes for their generation.

The US, Canada, Australian, Japan and other countries comply with Article 27 by including a broad generalist threshold requirement for patent eligible subject matter. In the US, for example section 101 of the Patent Act (35 U.S.C.) provides that: 'Whoever invents or discovers any new and useful process, machine, manufacture, or composition of matter, or any new and useful improvement thereof, may obtain a patent therefor, subject to the conditions and requirements of this title.'

Patent legislation in the US includes none of the available exclusions provided in Articles 27.2 and 27.3 of TRIPS. Member countries of the EPC and others have used a different approach, specifically listing the types of subject matter that do not constitute a patentable invention. Article 52 of the EPC, for example, excludes:

- discoveries, scientific theories and mathematical methods;
- aesthetic creations;
- schemes, rules and methods for performing mental acts, playing games or doing business, and programs for computers; and
- presentations of information.[27]

In addition to ordre public and morality, the EPC mirrors TRIPS with regard to the other Article 27 exclusions, save that the plant and animal exclusion only extends to plant and animal varieties. It should be noted at this point that although Brazil and some other South American countries take a similar approach to EPC members, they use somewhat different language. They also tend to include all of the specified TRIPS exclusions but, notably, add a further exclusion for biological materials. For example, Article 10 IX of the Brazilian Intellectual Property Law provides that the following are not to be considered inventions: 'natural living beings, in whole or in part, and biological material, including the genome or germ plasma of any natural living being, when found in nature or isolated therefrom, and natural biological processes'. The extent to which provisions of this nature comply with TRIPS requirement has not been tested.

The primary focus of this chapter is on recent United States case law and how it might affect the global agricultural biotechnology industry. One reason

---

26   The equivalent provision in the EPC was discussed in the preceding section of this chapter.

27   It should be noted that this provision excludes subject matter or activities 'only to the extent to which a European patent application or European patent relates to such subject matter or activities as such': Article 52(3) EPC.

for limiting the analysis in this chapter to the United States is that the interesting recent judicial pronouncements on patent eligibility of genetic technologies have occurred in this jurisdiction. But in addition, this is by far the largest jurisdiction in terms of the number of patents claiming genetic technologies, the level of financial investment in agricultural biotechnology, the number of biotechnology companies and the production of foodstuffs from genetically modified crops.[28] As such, patent decisions in the United States are likely to have the greatest impact globally. Even so, the global environment for patenting of genetic technologies can only be assessed fully by having regard to the laws in other countries and, more particularly, how those laws are interpreted by the judiciary.[29] Because judges in other countries are at liberty to interpret their own laws using approaches different from those of American judges, it would be unwise to read the impact of the recent United States decisions too broadly.

The Myriad Litigation

The litigation relating to Myriad's BRCA patents has followed a rather more complex path than most and events following the AMP SCOTUS decision are no less complex. An attempt is made in this section to explain the key events to make some sense of the take home messages for the genetic technology industry as a whole. A summary of the judicial decisions leading up to and including the AMP SCOTUS decision is presented in summary form in Table 6.1.[30]

Four types of patent claims were at issue in this case. The first claimed isolated DNA sequences that were identical to those existing naturally in the human genome, whether with or without mutated components. Some of the claims were to the entire genes, whereas others claimed much shorter sequences occurring within the genes. In addition to these claims to sequences that were isolated, but otherwise identical to their naturally occurring counterparts (commonly referred

---

28   See OECD, *Key biotechnology indicator* (last updated October 2013). [Online]. Available at: http://www.oecd.org/innovation/inno/keybiotechnologyindicators.htm [accessed 6 April 2014].

29   A recent Australian case illustrates this point. In *Cancer Voices v. Myriad Genetics, Inc* [2013] FCA 65, a case relating to the Australian equivalent of one of the patents in AMP SCOTUS, but decided before the SCOTUS decision, Nicholas J explained at [135] one of the reasons why he did not need to have regard to the US decision: 'the law in Australia is different. I must apply the law as explained in *NRDC*. It must also be recognised, especially as the [AMP] case heads to the US Supreme Court, that the constitutional setting in which patent legislation operates in the US is quite different to that in which patent legislation operates in this country'.

30   Hyperlinks to all of the decisions are available on a webpage compiled by the Centre for Public Genomics at Duke University. [Online]. Available at: http://www.genome.duke. edu/centers/cpg/BRCA-resources/ [accessed 6 April 2014]. For a more detailed analysis of the court decisions on patent eligible subject matter see Lawson, above n 10.

to as gDNA), a second set of claims were made to cDNA sequences. This type of DNA is made in the laboratory through a process of reverse transcription from messenger RNA. Messenger RNA is the naturally occurring intermediary in the process of transcription and translation for making proteins from genes. The genetic code for naturally occurring sequences (gDNA) is different from the code for cDNA because, in the process of making messenger RNA, parts of the code that are not required for protein synthesis are deleted. However, the function of the gDNA and cDNA derived from a particular gene is essentially the same, in that both are capable of coding for the same protein. The third and fourth sets of claims were to methods of using DNA sequences. The former claimed simple methods of comparing and analysing the relevant component of a patient's DNA sequence with a reference sequence, in order to detect the presence of mutations linked with increased risk of developing breast and/or ovarian cancer. The final single claim was to a more sophisticated method of screening for potential cancer therapeutics.

Table 6.1   Summary of decisions in the AMP case

	Isolated sequences identical to those in nature (gDNA)	Synthetic sequences complementary to RNA (cDNA)	Methods of comparing or analysing DNA sequences	Method for screening potential cancer therapeutics
AMP First Instance Sweet J	Invalid – products of nature. Physical embodiment of information – structural and functional differences irrelevant.	Not distinguished from gDNA.	Invalid – abstract mental processes. No machine or transformation.*	Invalid – basic scientific principle. No machine or transformation.
AMP Federal Circuit #1 and #2 Lourie J (majority)	Valid – markedly different chemical identity and nature – broken covalent bonds.	Valid – markedly different.	Invalid – abstract mental processes. Additional transformative steps were not included in the claims.	Valid – transformative. Growing and determining growth rates of transformed cells.
AMP Federal Circuit #1 and #2 Moore J	Valid – chemical differences are not enough and no new utility – not a blank canvas – leave intact settled expectations.	Valid – joined majority.	Invalid – joined majority.	Valid – joined majority.

	Isolated sequences identical to those in nature (gDNA)	Synthetic sequences complementary to RNA (cDNA)	Methods of comparing or analysing DNA sequences	Method for screening potential cancer therapeutics
AMP Federal Circuit #1 and #2 Bryson J	Invalid – products of nature – same structurally and functionally – the only changes were incidental to extraction.	Generally valid – agreed with the majority except for claims to short strands indistinguishable from gDNA.	Invalid – joined majority.	Valid – joined majority.
AMP SCOTUS	Invalid – products of nature. Isolation from surrounding genetic material is not enough.	Generally valid – markedly different, except for the claims identified by Bryson J.	Not decided.	Not decided.

Note: * It should be noted that Sweet J's decision was prior to SCOTUS decision in *Bilski v. Kappos*, 130 S. Ct. 3218 (2010). There, the majority ruled that the 'machine or transformation' test is only one factor in the consideration of the validity of process claims. The majority stated that while the test is a 'useful and important clue, an investigative tool, for determining whether some claimed inventions are processes under § 101', it is 'not the sole test' (at 3227). While this case has freed up the judiciary from the fetters of the machine or transformation test, it is submitted here that it would have been unlikely to have caused a change in Sweet J's decision on this point, based on the tenor of his other decision*s in this case.*

The court proceedings began with a motion by Myriad and other defendants to dismiss the complaint filed by AMP and the other complainants. Justice Sweet of the Federal District Court in the Southern District of New York (SDNY) denied the motion on 1 November 2009.[31] Subsequently, the complainants filed a motion for summary judgement, and on 29 March 2010 Sweet J handed down his decision that all four types of claims fail to satisfy the patent eligible subject matter requirement (AMP First Instance).[32] An appeal by the defendants to the Court of Appeals of the Federal Circuit followed. This court was given two opportunities to decide the case. The majority judgment in the first decision, handed down by Lourie J on 29 July 2011 reversed Sweet J's decision in part (AMP Federal Circuit #1).[33]

---

31   669 F. Supp. 2d 365 (SDNY 2009).

32   *Association for Molecular Pathology v. United States Patent and Trademark Office*, 702 F. Supp. 2d 181 (SDNY 2010).

33   *Association for Molecular Pathology v. United States Patent and Trademark Office*, 653 F.3d 1329 (Fed Cir 2011).

The decisions and reasoning of the other two judges who sat on the case are briefly summarised in Table 6.1.

Following lodgement of an appeal by AMP, the Supreme Court decided to refer the case back to the Federal Circuit to reconsider in light of an intervening decision in *Mayo Collaborative Services v. Prometheus Laboratories, Inc* (Mayo SCOTUS) concerning methods of diagnosis.[34] The Mayo SCOTUS decision was that, without more, a method of comparing and analysing rates of drug metabolism in the human body with reference data was patent ineligible on the basis that it amounted to patenting a law of nature. To be patent eligible, another inventive concept would have to be added, amounting to something 'significantly more than a patent upon the natural law itself'.[35] The Court was also concerned that claims relating to laws of nature could result in 'tying up the use of the underlying natural laws, inhibiting their use in the making of further discoveries'.[36] Such claims are said impermissibly to preempt use of laws of nature.[37]

The Federal Circuit was asked to reconsider the decision in AMP Federal Circuit #1 in light of Mayo. It should be noted at this point that the court had already invalidated most of Myriad's method claims and it is difficult to see how Mayo SCOTUS might have affected the reasoning of judges in relation to the sequence claims. Taking these factors into account, it probably comes as no surprise that, on remand, the Federal Circuit's decision in AMP remained largely unchanged (AMP Federal Circuit #2).[38] The complainants then made a further appeal to the Supreme Court. The appeal was limited to the first and second sets of claims, which the majority of the Federal Circuit had previously held (twice) were valid. The final SCOTUS decision was handed down on 13 June 2013 (AMP SCOTUS).[39] The court was unanimous in holding that the first set of claims to gDNA failed to satisfy the patent eligible subject matter requirement, but that the second set of claims to cDNA were valid.

In a nutshell, the Supreme Court decided that 'a naturally occurring DNA segment is a product of nature and not patent eligible merely because it has been isolated, but ... cDNA is patent eligible because it is not naturally occurring.'[40] The Court held that gDNA fails to satisfy the Chakrabarty requirement of exhibiting markedly different characteristics from any found in nature, stating that '[t]o be sure, [Myriad] found an important and useful gene, but separating that gene from

---

34   *Mayo Collaborative Services v. Prometheus Laboratories, Inc*, 132 S. Ct. 1289 (2012).

35   Ibid., 1294.

36   Ibid.

37   For a critique on the Supreme Court's approach to preemption in Mayo and other cases see: Strandburg, K. 2013. Much ado about preemption. *Houston Law Review*, 50(2), 563–622, describing the preemption rhetoric as a 'red herring' (at 566).

38   *Association for Molecular Pathology v. United States Patent and Trademark Office*, 689 F.3d 1303, 1337 (Fed Cir 2012).

39   *Association for Molecular Pathology v. Myriad Genetics, Inc.*, 133 S. Ct. 2107 (2013).

40   Ibid., 2111.

its surrounding genetic material is not an act of invention.'[41] In coming to this decision the Court rejected the rationale for upholding the validity of these gDNA claims posited by Lourie J in the Federal Circuit decisions, which focused on the chemical rather than the informational nature of DNA.[42] Justice Lourie held that breaking of the covalent bonds to isolate a gene from the rest of the DNA molecule was enough to make the isolated gene a human made invention with markedly different characteristics from any found in nature.[43] In contrast, the Supreme Court concluded that, because the claims were concerned primarily with the information encoded in the BRCA genes rather than the composition of the molecule, Myriad could not rely on the chemical differences between the naturally occurring and isolated forms of the sequences.[44]

Reflections on the Implications of the AMP SCOTUS Decision

Since the AMP SCOTUS decision was handed down, a number of commentators in the United States and elsewhere have tried to make sense of the decision itself and its broader implications, both in online blogs[45] and in academic publications.[46] The holding itself seems clear. As noted by Charles Lawson:

> The Supreme Court's decision has undoubtedly clarified the un-patentability of claims to naturally occurring DNA in the form of '[t]he isolated DNA of claim 1, wherein said DNA has the nucleotide sequence set forth in SEQ ID NO:1'

---

41   Ibid., 2116.

42   Ibid., 2118.

43   *Association for Molecular Pathology v. United States Patent and Trademark Office*, 653 F.3d 1329, 1352 (Fed Cir 2011); *Association for Molecular Pathology v. United States Patent and Trademark Office*, 689 F.3d 1303, 1355 (Fed Cir 2012).

44   *Association for Molecular Pathology v. Myriad Genetics, Inc.*, 133 S. Ct. 2107, 2118 (2013).

45   For example: Genomics Law Report, *Myriad Gene Patent Litigation*. [Online]. Available at: http://www.genomicslawreport.com/index.php/category/badges/myriad-gene-patent-litigation/ [accessed 14 April 2014]; Patent Docs, *Supreme Court Issues Decision in AMP v. Myriad*. [Online]. Available at: http://www.patentdocs.org/2013/06/supreme-court-issues-decision-in-amp-v-myriad.html [accessed 14 April 2014].

46   For example: Burk, D. 2014. The curious incident of the Supreme Court in *Myriad Genetics*. *Notre Dame Law Review*, 90, in press; Holman, C. 2013. Editorial: In Myriad the Supreme Court has, once again, increased the uncertainty of U.S. patent law. *Biotechnology Law* Report, 32(5), 289–93; Lawson, above n 10; Waller, P. and Young, D. 2014. The new world of diagnostic testing post-*Myriad* (2014) 33 *Biotechnology Law Report*, 33(1), 17–18; Rai, A. 2013. Biomedical patents at the Supreme Court: A path forward. *Stanford Law Review Online*, 66, 111–16; Olson, D. 2013. Patent protection for genetic innovation: Monsanto and Myriad. *Cato Supreme Court Review*, 283–300. [Online]. Available at: http://object.cato.org/sites/cato.org/files/serials/files/supreme-court-review/2013/9/olson.pdf [accessed 15 May 2014].

(the '282 patent, claim 2) and '[a]n isolated DNA having at least 15 nucleotides of the DNA of [the nucleotide sequence set forth in SEQ ID NO:1]' (the '282 patent, claim 6).[47]

The Supreme Court's rationale for distinguishing between gDNA and cDNA is somewhat difficult to understand if the key attribute of DNA that remains unchanged from its natural to its isolated gDNA form is its informational content. Admittedly there are clear differences in informational content between naturally occurring and cDNA, because non-coding parts of the naturally occurring sequence are omitted from the cDNA sequence. However, those differences do not affect the function of coding for particular proteins. Thus, the *functional information* carried in the gDNA sequence and cDNA sequence is identical.[48] The marked difference in Chakrabarty was one the Supreme Court held had 'the potential for significant utility',[49] suggesting that in the circumstances of that case the difference was functional. It is less clear whether it is an absolute requirement of the marked difference test that the difference is functional. If there is a need for a functional difference, might there be other characteristics of cDNA that make it markedly different? In invalidating the gDNA claims, the Supreme Court focused on their quality of being isolated. Arguably, one point of distinction for cDNA is that it is synthetically manufactured in the laboratory. The capacity to make synthetic DNA in the laboratory would clearly have some utility, in both research and clinical contexts. Could it be that other synthetic claims might also be valid, using this logic? While an interlocutory District Court decision (discussed below) suggests that this may not be the case, Chris Holman notes that the SCOTUS decision itself is 'not altogether unambiguous on this point'.[50] Nevertheless, even he concludes that the structural differences seem more likely to be critical to the Supreme Court's holding that cDNA is patent eligible but gDNA is not.[51]

There is as much uncertainty over the broader implications of the AMP SCOTUS decision as there is in the rationale for the decision itself. Some commentators have suggested that, despite the somewhat convoluted reasoning, the Supreme Court achieved the right outcome. For example, David Olson posits that:

> Although the Court in Myriad did not adopt the most straightforward and efficient way of determining patent eligible subject matter, its parsing of patentability for DNA and cDNA may have served the function of granting enough patent rights to incentivize research, but not so many patents as to cause very high consumer costs and research blockages. By preserving naturally occurring DNA in the public domain, the Court made sure this material is free of patents and available to all.

---

47   Lawson, ibid., 15.
48   On this point see Rai, above n 47; Burk, above n 47.
49   *Diamond v. Chakrabarty,* 447 US 303, 310 (1980).
50   Holman, above n 47, 290–91.
51   Ibid., 292.

By allowing cDNA to be patented, the Court ensured that some incentives flow to
genetic researchers for their discoveries of important gene disease correlations.[52]

This and other commentary hints at the possibility that future courts and the
USPTO will give the decision a narrow reading, restricting it to the broad sequence
claims found in the Myriad patents and other similar patents granted in the 1980s
and 1990s in the early developmental phase of the genetic technology.[53] Others
are much less circumspect, suggesting that the patent eligibility of a large range of
chemical compounds could be open to question.[54]

One omission from much of the commentary to date is an evaluation of the
consequences for patents relating to non-human genetic technologies.[55] Although
the subject matter in issue in the AMP litigation was human gDNA and cDNA,
it is difficult to imagine that the ruling could be interpreted as being restricted to
human DNA.[56] There are no sufficiently distinctive differences in the chemical or
informational content of DNA between species to justify such an interpretation.
As such, while the AMP SCOTUS decision may have freed the genetic diagnostics
sector from some of the concerns about the negative aspects of gene patenting, it
may well have created a different outcome for other genetic technologies.

The AMP SCOTUS ruling on the failure of the BRCA gDNA claims to satisfy
the patent eligible subject matter requirement thus raises a gamut of questions,
including the following:

- does the ruling on patent ineligibility apply only to claims to isolated
  gDNA or does it extend to synthetically produced DNA that is not markedly
  different from natural DNA;
- is the ruling restricted to human DNA, or does it extend to all products
  isolated from the natural world; and
- how much difference must exist for an isolated product to be 'markedly
  different' from its naturally occurring counterpart?

Depending on the answers to such questions, the decision could create only a
minor hiccup for the genetic technology industry, invalidating old, broad patents

---

52 Olson, above n 47.

53 See, for example, Lawson, above n 10, 5.

54 See, for example, Rai, above n 47; Burk, above n 47.

55 Noting, however, that some commentators do specifically mention the implications
for agricultural biotechnology. See, for example, Holman, above n 47, 290.

56 Chris Holman confirms this view: ibid., 290. Writing before the AMP SCOTUS
decision was handed down, Greg Graff and colleagues noted that if the court were to make
such a distinction between human and non-human DNA, '[u]nder such a biologically split
decision, it would be necessary for the Court to provide clear rules for differentiating between
(unpatentable) human sequences and (patentable) non-human sequences: Graff, G., Phillips, D.,
Lei, Z., Oh, S., Nottenburg, C. and Pardey, P. 2013. Not quite a Myriad of gene patents. *Nature
Biotechnology*, 31(5), 404–10, 406. As we now know, that clear distinction did not eventuate.

that are nearing expiry and requiring small adjustments in future claims drafting. Alternatively, the decision could significantly impact on the availability of patents for any subject matter that, in some way or another, is linked to the natural world.

The earlier Federal Circuit AMP decisions invalidating most of Myriad's method claims also need to be taken into account. Although the Supreme Court did not have the opportunity to reconsider these claims, its earlier jurisprudence, particularly the Mayo SCOTUS decision, raises an additional set of questions. It should be recalled at this point that the Supreme Court in Mayo held that method claims relating to laws of nature, natural phenomena, or abstract ideas are only patent eligible if they incorporate another inventive concept.[57] The key questions in relation to methods relating to genetic technologies are whether products of nature are seen in the same light as laws of nature[58] and what level of inventive concept is required to transform such methods into patent eligible subject matter. The question of preemption also needs to be considered, particularly whether the Mayo SCOTUS ruling on preemption of laws of nature applies in the same way to products of nature.

Sequelae to the AMP SCOTUS Decision

Recent events suggest that a broad approach to the interpretation of the AMP SCOTUS decision and the Mayo decision may be favoured by both the judiciary and the USPTO. If this is the case, then the impact may be significant and is likely to be felt in the agricultural sector just as much as in biomedicine.

The Judiciary

Following the SCOTUS decision, Ambry and Gene-by-Gene, two of Myriad's closest competitors, announced that they would start offering BRCA diagnostic genetic tests.[59] In July 2013, Myriad instituted proceedings in the US District Court of New York alleging that both companies were infringing a number of its remaining patents, which cover a range of claims to methods of diagnosis and to synthetically produced DNA sequences.[60] Myriad has since instituted further infringement proceedings against other competitors.[61]

---

57   *Mayo Collaborative Services v. Prometheus Laboratories, Inc*, 132 S. Ct. 1289, 1294 (2012).

58   See Burk, above n 47.

59   Cook-Deegan, R. 2014. *Viewers' guide to the Super Bowl of gene patent cases. The Cancer Letter* (14 March 2014) < http://www.cancerletter.com/articles/201403144>.

60   Noonan, K. 2013. Myriad Genetics files suit against Ambry Genetics for genetic diagnostic testing of BRCA genes. *Patent Docs* (9 July). [Online]. Available at: http://www.patentdocs.org/2013/07/myriad-genetics-files-suit-against-ambry-genetics-for-genetic-diagnostic-testing-of-brca-genes.html [accessed 20 May 2014]; Cook-Deegan, ibid.

61   Cook-Deegan, ibid.

In September 2013, Myriad applied for a preliminary injunction against Ambry and Gene-by-Gene, requiring those companies to cease offering BRCA and other tests until the case was decided. Shelby J handed down his decision refusing Myriad's application on 10 March 2014.[62] The claims in issue here can usefully be separated out into two types.[63] The first type of claims are to primers, short strands of DNA which are used in the process of amplification (making multiple copies) of naturally occurring BRCA genes in the laboratory. Amplification is a necessary part of genetic testing methodology. As pointed out by Shelby J, the primers must be complementary to the relevant segment of the naturally occurring BRCA genes to be capable of binding to them.[64] The second type of claims are to the methods of comparing reference BRCA sequences with patient sequences using gene probes or assays.[65] Gene probes are essentially the same as primers, except that they usually have some sort of marker attached to them, like a fluorescent dye.

Recognising that it is difficult to do justice to the entire 106 page judgement here, what follows is a brief overview of the key aspects of the decision. Although Shelby J accepted that the plaintiffs would suffer irreparable harm if the injunction did not issue,[66] he did not accept that they were likely to succeed on the merits in respect of either type of claim, and refused the injunction on this basis. The plaintiffs tried to argue that because the primers were made in the laboratory and were synthetic, they satisfied the subject matter requirement, arguing that the Supreme Court in AMP found only extracted, genomic DNA to be patent ineligible.[67] Shelby J rejected this argument, in part based on the following statement in AMP SCOTUS: 'cDNA is not a 'product of nature' and is patent-eligible under §101, except insofar as very short series of DNA may have no intervening introns to remove when creating cDNA. In that situation, a short strand of cDNA may be indistinguishable from natural DNA.'[68] Shelby J extrapolated that: 'If cDNA – which is clearly synthetic – is sometimes patent ineligible, then implicit in the Supreme Court's decision is the conclusion that not all synthetic DNA is patent eligible.'[69] This clearly signals that it is not the

---

62 *In re BRCA1- and BRCA 2-Based Hereditary Cancer Test Patent Litigation: Memorandum Decision and Order Denying Plaintiffs' Motion for Preliminary Injunction*, Case No 2:13-CV-00640-RJS (DUCD 2014).

63 *In re BRCA1- and BRCA 2-Based Hereditary Cancer Test Patent Litigation: Memorandum Decision and Order Denying Plaintiffs' Motion for Preliminary Injunction*, Case No 2:13-CV-00640-RJS (DUCD 2014) slip opinion, 13–14; 48–54.

64 Ibid.

65 See generally ibid., 13–16, 48–54.

66 Ibid., 62 (price erosion and loss of market share) and 66 (loss of exclusivity of patent term); but note 65 (not reputational harm).

67 Ibid., 75.

68 *Association for Molecular Pathology v. Myriad Genetics, Inc*, 133 S. Ct. 2107, 2119 (2013). See also *In re BRCA1- and BRCA 2-Based Hereditary Cancer Test Patent Litigation*, above n 64, 79.

69 *In re BRCA1- and BRCA 2-Based Hereditary Cancer Test Patent Litigation*, ibid., 80.

synthetic nature of cDNA that makes it patent eligible but the difference in informational content, at least in the eyes of Shelby J.

Myriad's method claims suffered the same fate. Here Shelby J relied on the decision in Mayo SCOTUS, drawing an analogy between the products of nature that formed the basis of the method claims in this case with the laws of nature in Mayo SCOTUS.[70] His Honour concluded that the addition of hybridised gene probes or assays to the method claims that were previously held to be patent ineligible in both of the AMP Federal Circuit decisions were well-understood, routine, conventional activities. Hence they were patent ineligible in accordance with Mayo SCOTUS.[71] His Honour also accepted the argument that these claims could be preemptive, relying on a recent case also relating to methods of using DNA, which the Federal Circuit held were patent ineligible because they preempted all practical uses of a product of nature.[72]

Shelby J's holding on the first type of claims is perhaps the least controversial aspect of his decision.[73] The entire purpose of primers and probes is to hybridise to naturally occurring segments of DNA and hence they are required to be identical. The comment in the AMP SCOTUS decision about patent ineligibility of short strands of cDNA that are indistinguishable from natural DNA left him little room to manoeuvre. However, the decision on method claims is more controversial and could be more far reaching. It seems to be another step along the path taken in AMP SCOTUS that, according to Dan Burk, 'seems to elide any distinction between the two categories [of laws of nature and products of nature]'.[74] In closing this discussion of the decision by Shelby J, it should be noted that this is only a first instance decision and, given the course of litigation relating to the BRCA patents to date, an appeal seems likely.

The USPTO

On 4 March 2014 the USPTO released a guidance document for patent examiners when determining subject matter eligibility of claims relating to the judicial exceptions of laws of nature, natural phenomena and natural products.[75] The exception for abstract ideas is dealt with through a separate guidance

---

70   Ibid., 93.

71   Ibid., 94–5.

72   *Aria Diagnostics, Inc. v. Sequenom, Inc.*, 726 F.3d 1296 (Federal Circuit 2013). See *In re BRCA1- and BRCA 2-Based Hereditary Cancer Test Patent Litigation*, above n 64, 96–100.

73   Noting that there is some disagreement on this point.

74   Burk, above n 47, 8.

75   USPTO. 2014. *Guidance for determining subject matter eligibility of claims reciting or involving laws of nature, natural phenomena and natural products*. [Online]. Available at: http://www.uspto.gov/patents/law/exam/myriad-mayo_guidance. pdf [accessed 15 April 2014].

document,[76] which it is not necessary to elaborate on further for the purpose of this chapter. The guidance document sets out three questions that examiners must address to determine patent eligibility. The third question is the key one for present purposes. It asks whether 'the claim as a whole recite[s] something *significantly different* than the judicial exception(s)'.[77] The guidance document lists the type of subject matter that must be analysed under this question, which, it states:

> includes, but is not limited to: chemicals derived from natural sources (e.g., antibiotics, fats, oils, petroleum derivatives, resins, toxins, etc.); foods (e.g., fruits, grains, meats and vegetables); metals and metallic compounds that exist in nature; minerals; natural materials (e.g., rocks, sands, soils); nucleic acids; organisms (e.g., bacteria, plants and multicellular animals); proteins and peptides; and other substances found in or derived from nature ... This is the case regardless of whether particular words (e.g., 'isolated', 'recombinant', or 'synthetic') are recited in the claim.[78]

The guidance document then lists the types of factors that weigh towards eligibility and other, largely mirror-image factors that weigh against eligibility. Relevant considerations include whether: a product claim that appears to claim a natural product is determined, following analysis, to be non-naturally occurring and markedly different in structure; there are meaningful limits on claim scope to avoid substantially foreclosing the judicial exception; there are elements or steps in addition to the judicial exception that relate to it in a significant way; there are elements or steps in addition to the judicial exception that more than merely describe it; there are elements or steps in addition to the judicial exception that include a machine or transformation; and there are elements or steps that add a feature 'that is more than well-understood, purely conventional or routine in the relevant field'.[79] Eight examples are provided to assist examiners further, some of which are drawn from cases like *Chakrabarty* (plasmids – not eligible; bacterium with plasmids – eligible),[80] *Funk Bros* (inoculant for leguminous plants comprising a combination of strains of bacteria – not eligible)[81] and AMP SCOTUS (primers – not eligible; method claims using the primers with an additional step of heating and cooling the reaction mixture – eligible).[82] Another example indicates that a

---

76   USPTO. 2014. *Manual of Patent Examining Procedure*, section 2106. [Online]. Available at: http://www.uspto.gov/web/offices/pac/mpep/s2106.html [accessed 15 April 2014].

77   USPTO, above n 76, 3 [emphasis in original].

78   Ibid. [emphasis in original].

79   Ibid., 4–5.

80   Ibid., Example A, 5–7.

81   *Funk Brothers Seed Co. v. Kalo Inoculant Co.*, 333 U.S. 127 (1948). USPTO, ibid., Example D, 10–11.

82   USPTO, ibid., Example H, 18.

purified naturally occurring chemical will not be eligible, even where there is a clear material advantage and many others had tried and failed before the applicant succeeded.[83] However, a method of applying the purified product may be patentable if, for example, the claim is limited to specified dosages and time frames.[84]

In sum, this guidance document clearly demonstrates that in the eyes of the USPTO the holding in AMP SCOTUS extends far beyond human gene sequences to any product derived from natural sources. In this regard, the guidance provided to patent examiners is more far reaching than strictly required from the AMP SCOTUS decision. By including all products from the natural world the guidance document appears to ignore the most significant problem the Supreme Court had with the gDNA BRCA claims – that the claims related to the distinctive informational nature of DNA, but the alleged marked differences were only in respect of its chemical structure. Lawson puts forward the suggestion that 'if the Myriad claims had been in the form of chemical composition claims they might have been patent eligible'.[85] However, under the new USPTO guidance document it seems unlikely that a patent examiner would agree.

In other respects, however, the guidance document is more circumspect and appears to interpret the requirement for a marked difference in what could be described as a minimalist way, as illustrated by the above examples and by the following extract:

> The fact that a marked difference came about as a result of routine activity or via human manipulation of natural processes does not prevent the marked difference from weighing in favor of patent eligibility. For example, cDNA having a nucleotide sequence that is markedly different from naturally occurring DNA is eligible subject matter, even though the process of making cDNA is routine in the biotechnology art. *See [AMP SCOTUS]* at 2119. Similarly, a hybrid plant that is markedly different from naturally occurring plants is eligible subject matter, even though it was created via routine manipulation of natural processes such as pollination and fertilization. *See, e.g., J.E.M. Ag Supply, Inc. v. Pioneer Hi-Bred Int'l, Inc.*, 534 U.S. 124, 145 (2001).[86]

Quite how the guidance document will affect future patent applications relating to products and processes connected with the natural world remains to be seen. Some commentators have been quick to point out that the guidance document is what it says it is, a mere guideline which is (and doubtless will be) subject to challenge in the courts.[87]

---

83   See ibid., Example B, 7–9.

84   Ibid.

85   Lawson, above n 10, 14.

86   USPTO, above n 76, 5.

87   See, for example, Noonan, K. 2014. Thoughts on the USPTO's patent eligibility guidelines (and what to do about them), *Patent Docs* March 18 2014. [Online]. Available

Concluding Remarks: What Does This Means for Agricultural Biotechnology?

It is difficult to state with any certainty what this means for agricultural biotechnology. There is considerable uncertainty in distinguishing between patent eligible and patent ineligible subject matter in this field following AMP SCOTUS and the Mayo SCOTUS decision. The USPTO guidance document may have added to this uncertainty, rather than alleviating it.

It does seem clear that there can be no good reason in logic or precedent to distinguish between human and non-human DNA. Hence the ruling in AMP SCOTUS that isolated DNA which is identical to its natural counterpart is patent ineligible would in all likelihood apply in the agricultural field as well as in medicine. Greg Graff and colleagues recently analysed the number of US patents that include claims to DNA sequences, using a combination of methods including sophisticated linguistic algorithms.[88] Although their analysis was undertaken prior to the AMP SCOTUS decision, they speculated on the number of patent claims that would be invalidated if the Federal Circuit decisions relating to patent eligibility of isolated gDNA were overturned. They calculated that '12% of the total 72,052 nucleotide-related patents found in the US patent literature have claims at risk of invalidation,' and that more than half of these protect applications in fields such as 'veterinary medicine, crop agriculture, food and beverage manufacturing, industrial enzymes or bioenergy'.[89] Despite this, they conclude that the outcome of the AMP SCOTUS decision may be less profound than feared on the basis that, for more than a decade, there has been a move away from claiming simple isolated gDNA, even though the number of patents claiming DNA sequences continues to grow.[90]

With regard to other subject matter, the USPTO guidance document lists multicellular organisms as subject matter that is potentially patent ineligible. Despite this, it would be difficult to argue that GMOs, which are clearly markedly different from naturally occurring equivalents, would be patent ineligible as a result of the SCOTUS AMP decision.[91] Does this mean that patent holders in the

---

at: http://www.patentdocs.org/2014/03/thoughts-on-the-usptos-patent-eligibility-guidelines-and-what-to-do-about-them.html [accessed 15 April 2014].

88   Graff *et al.*, above n 57.

89   Ibid., 407.

90   Ibid., 408.

91   If a decision along the lines of AMP SCOTUS were handed down in Canada, there would be more significant consequences at the level of the whole organism in that jurisdiction. In *Harvard College v. Canada (Commissioner of Patents)* [2002] 4 SCR 45, the Canadian Supreme Court held that genetically modified higher organisms were not patentable. However, two years later in *Monsanto Canada Inc. v. Schmeiser* [2004] 1 SCR 902 the Supreme Court essentially gave *de facto* patent protection for genetically modified organisms by recognising the patent eligibility of genes and cells. Note also that on 8 May 2014 the United States Court of Appeals of the Federal Circuit rejected an appeal from a decision of the USPTO refusing to allow a patent for cloned animals produced by

agricultural biotechnology industry can rest easy, largely ignoring the brouhaha around *AMP* and *Mayo*? That would probably be unwise. As noted at the start of this chapter, there is a vast array of other methods and products that are patented in the agricultural biotechnology sector. The injunction decision by Shelby J and, more particularly, the new USPTO guidance document, indicate that many of these could now be open to challenge on the basis that they may include claims that are patent ineligible. In one way or another, all are connected to the natural world. What threshold product claims would need to cross in order to be markedly different is not yet clear. The additional elements or steps that need to be included in method claims are no clearer. Although the USPTO has made efforts to provide examples of how these matters might be resolved, until these questions are decided by the courts uncertainty will remain. Indeed, as Kevin Noonan points out, the only decisions that really matter are those of the Supreme Court.[92]

the 'Dolly the sheep' technique: *In Re Roslin Institute (Edinburgh)* Case No 2013–1407. The basis for the decision was that the clones are not markedly different from their naturally occurring counterparts, following *Chakrabarty, AMP SCOTUS* and other decisions.

92   Noonan, K. 2012. The Supreme Court's *Prometheus* decision and its meaning for diagnostic method claims. *Biotechnology Law Report*, 31(3), 267–70, 268.

Chapter 7

Just Label It: Consumer Rights, GM Food Labelling and International Trade

Matthew Rimmer

Introduction

In the US, there has been fierce debate over state, federal and international efforts to engage in GM food labelling. A grassroots coalition of consumers, environmentalists, organic farmers and the food movement has pushed for law reform in respect of GM food labelling. The *Just Label It* campaign has encouraged US consumers to send comments to the US Food and Drug Administration (FDA) to label GM foods.[1]

This chapter explores the various justifications made in respect of GM food labelling. There has been a considerable effort to portray the issue of GM food labelling as one of consumer rights as part of 'the right to know'. There has been a significant battle amongst farmers over GM food labelling – with organic farmers and biotechnology companies, fighting for precedence. There has also been a significant discussion about the use of GM food labelling as a form of environmental legislation. The prescriptions in GM food labelling regulations may serve to promote eco-labelling and deter greenwashing.[2] There has been a significant debate over whether GM food labelling may serve to regulate corporations – particularly from the food, agriculture and biotechnology industries. There are significant issues about the interaction between intellectual property laws – particularly in respect of trade mark law and consumer protection – and regulatory proposals focused upon biotechnology. There has been a lack of international harmonisation in respect of GM food labelling. As such, there has been a major use of comparative arguments about regulator models in respect of food labelling. There has also been a discussion about international law, particularly with the emergence of sweeping regional trade proposals, such as the *Trans-Pacific Partnership* and the *Trans-Atlantic Trade and Investment Partnership*.

This Chapter considers the US debates over GM food labelling – at state, federal, and international levels. The battles often involved the use of citizen-

---

1 *Just Label It*. [Online]. Available at: http://www.justlabelit.org [accessed 2 April 2014].

2 Lane, E. 2011. *Clean Tech Intellectual Property: Eco-Marks, Green Patents, and Green Innovation*, Oxford: Oxford University Press.

initiated referenda. The policy conflicts have been policy-centric disputes – pitting organic farmers, consumers and environmentalists against the food industry and biotechnology industry. Such battles have raised questions about consumer rights, public health, freedom of speech and corporate rights. The disputes highlighted larger issues about lobbying, fund-raising and political influence. The role of money in the US has been a prominent concern of Lawrence Lessig in his recent academic and policy work with the group Rootstrikers.[3] Part 1 considers the debate in California over Proposition 37. Part 2 explores other key state initiatives in respect of GM food labelling. Part 3 examines the Federal debate in the US over GM food labelling. Part 4 explores whether regional trade agreements – such as the *Trans-Pacific Partnership (TPP)* and the *Trans-Atlantic Trade and Investment Partnership (TTIP)* – will impact upon initiatives in respect of GM food labelling.

Proposition 37: The California Right to Know Genetically Engineered Food Act

In 2012 there was an intense debate in California over Proposition 37.[4] Proposition 37 would require the labelling of food made from plants or animals with genetic material changed in specified ways and sold to consumers. Proposition 37 would also prohibit the marketing of such food, or other processed food, as 'natural'. The regime has a number of exceptions and limitations – including foods that are:

> certified organic; unintentionally produced with genetically engineered material; made from animals fed or injected with genetically engineered material but not genetically engineered themselves; processed with or containing only small amounts of genetically engineered ingredients; administered for treatment of medical conditions; sold for immediate consumption such as in a restaurant; or alcoholic beverages.[5]

The advocates of Proposition 37[6] argue that the measure would 'give us the right to know what is in the food we eat and feed to our families'[7] and require labelling of food produced using genetic engineering, so we can choose whether to buy

---

3   Lessig, L. 2011. *Republic, Lost: How Money Corrupts Congress and a Plan to Stop It, Twelve Books*, 2011; Lessig, L. 2012. *One Way Forward: The Outsider's Guide to Fixing the Republic*, eBook, 2012; and Lessig, L. 2013. Lesterland: *The Corruption of Congress and How to End It*, TED Books, April 2013.

4   *California General Election, Proposition 37.* [Online]. Available at: http://voterguide.sos.ca.gov/past/2012/general/propositions/37/ [accessed 29 March 2014].

5   Attorney General, *Proposition 37: Genetically Engineered Foods. Labeling. Initiative Statute.* [Online]. Available at: http://repository.uchastings.edu/cgi/viewcontent.cgi?article=2315&context=ca_ballot_props [accessed 2 April 2014].

6   See *California Right to Know – Label Genetically Engineered Foods.* [Online]. Available at: http://www.carighttoknow.org/ [accessed 2 April 2014].

7   *California General Election, Proposition 37,* above n 4.

those products or not'[8] because 'we have a right to know.'[9] The advocates of the measure maintained that:

> Fifty countries around the world – representing more than 40 per cent of the world's population – already require GMO labelling, including all of Europe, Japan, India and China. Polls show that more than 90 per cent of Americans want to know if their food is genetically engineered. We are free to choose what we want to eat and feed our children. The free market is supposed to provide consumers with accurate information about products so we can make informed choices.[10]

To support its proposition, the 'Yes' Campaign enlisted a number of celebrities to provide endorsements in an advertising campaign. An all-star cast of Hollywood actors appeared in a piece on the 'Right to Know'.[11] The actor and comedian Danny DeVito led the piece, asking the question: 'What makes you think you have the right to know?'[12] He joked: 'Knowing if you're eating or buying genetically engineered food is not your right.'[13] This montage also featured spokespersons Bill Maher, Dave Matthews, Jillian Michaels, Emily Deschanel, John Cho, Glenn Howerton, Kaitlin Olson, KaDee Strickland and Kristin Bauer van Straten. The video featured the affirmation – 'Demand that GMO's are labelled.' This witty video was designed to garner support for Proposition 37 by the consumer advocacy group Food and Water Watch.[14]

Marisa Tomei and other celebrities appeared in a video, emphasising that GM food labelling would not increase grocery costs.[15] Another video entitled 'I'm a Mom' – featured celebrities like Molly Ringwald, attesting 'I'm a Mom: if you label it, then I'll Know.'[16] James Franco also fronted a video, encouraging

---

8   Ibid.

9   Ibid.

10   California Right to Know. *Facts – Yes on Prop 37*. [Online]. Available at: http://www.carighttoknow.org/facts [accessed 2 April 2014].

11   Gates, S, 2012. California Prop 37: Danny DeVito, Dave Matthews and other celebs ask voters to approve mandatory GMO labelling. *The Huffington Post*, 9 October 2012. [Online]. Available at: http://www.huffingtonpost.com/2012/10/09/california-prop-37-video-celebrities_n_1952203.html [accessed 2 April 2014].

12   Food and Water Watch. 2012. *Right to know: Vote yes on Prop 37*, YouTube, 8 October 2012. [Online]. Available at: https://www.youtube.com/watch?v=RB1xHFwSYIg [accessed 2 April 2014].

13   Ibid.

14   Gates, above n 11.

15   Food and Water Watch. 2012. *Grocery costs: Vote yes on Pro 37*, YouTube, 2 November 2012. [Online]. Available at: https://www.youtube.com/watch?feature=player_embedded&v=i7V3q2HDQLI [accessed 2 April 2014].

16   Food and Water Watch. 2012. *I'm a mom: Vote yes on Prop 37*, YouTube, 2 November 2012. [Online]. Available at: https://www.youtube.com/watch?feature=player_embedded&v=7XcAcFTjm0g [accessed 2 April 2014].

Californian voters to vote 'Yes.'[17] In another video, James Franco appeared in a blindfold, observing, 'Right now, you are eating food with a blindfold on … because companies do not have to tell you whether the food has been genetically modified.'[18]

Stacy Malkan, an advocate for environmental health, was a Yes spokesperson on the *37 California Right to Know Campaign*.[19] She commented:

> Proposition 37 is very simple. It's about our right to know what's in the food we're eating and feeding our families. It's about our right to decide if we want to eat food that's been fundamentally altered at the genetic level, by companies like Monsanto, to contain bacteria, viruses or foreign genes that have never been in the food system before. And genetic engineering has been hidden from American consumers for two decades. Sixty-one other countries require labelling laws, but we haven't been able to get labelling here because of the enormous influence of Monsanto and the chemical companies.[20]

Malkan maintained that the proposal was supported by leading environmental, labour, and consumer groups in California. Malkan argued that the opponents of GM labelling were running an astroturfed, faux populist campaign against Proposition 37: 'On the other side are the world's largest pesticide and junk food companies, who are spending $40 million carpet-bombing California with a campaign of deception and trickery, with lie after lie in the ads that are going unchallenged in the media.'[21]

Writing in *The New York Times*, Michael Pollan wondered whether the push for GM food labelling was part of a larger food movement.[22] He commented that the battle over Proposition 37 raised larger issues about 'Big Food' and the industrial production of food:

> What is at stake this time around is not just the fate of genetically modified crops but the public's confidence in the industrial food chain. That system is

---

17   Food and Water Watch. 2012. *Grocery costs, featuring James Franco: Vote Yes on Prop 37*, YouTube, 2 November 2012. [Online]. Available at: https://www.youtube.com/watch?feature=player_embedded&v=pxeWHvJJS54 [accessed 2 April 2014].

18   Food and Water Watch. 2012. *Blindfold: Vote Yes on Prop 37*, YouTube, 2 November 2012. [Online]. Available at: https://www.youtube.com/watch?feature=player_ embedded&v=MFAii2gGdjg [accessed 2 April 2014].

19   Goodman, A. 2012. Food Fight: Debating Prop 37, California's Landmark Initiative to Label GMO Food. *Democracy Now*, 24 October 2012. [Online]. Available at: http://www.democracynow.org/2012/10/24/food_fight_debating_prop_37_californias [accessed 2 April 2014].

20   Ibid.

21   Ibid.

22   Pollan, M. 2012. Vote for the dinner party. *The New York Times*, 10 October 2012. [Online]. Available at: http://www.nytimes.com/2012/10/14/magazine/why-californias-proposition-37-should-matter-to-anyone-who-cares-about-food.html?pagewanted=all&_r=0 [accessed 2 April 2014].

being challenged on a great many fronts – indeed, seemingly everywhere but in Washington. Around the country, dozens of proposals to tax and regulate soda have put the beverage industry on the defensive, forcing it to play a very expensive (and thus far successful) game of Whac-A-Mole. The meat industry is getting it from all sides: animal rights advocates seeking to expose its brutality; public-health advocates campaigning against antibiotics in animal feed; environmentalists highlighting factory farming's contribution to climate change.

Pollan noted: 'The industry is happy to boast about genetically engineered crops in the elite precincts of the op.-ed and business pages – as a technology needed to feed the world, combat climate change, solve Africa's problems, etc. – but still would rather not mention it to the consumers who actually eat the stuff.'[23] He observed that such a lack of transparency was contradictory, maintaining that 'the fight over labelling GM food is not foremost about food safety or environmental harm, legitimate though these questions are.'[24] He stressed: 'The fight is about the power of Big Food.'[25] Pollan observed that:

Monsanto has become the symbol of everything people dislike about industrial agriculture: corporate control of the regulatory process; lack of transparency (for consumers) and lack of choice (for farmers); an intensifying rain of pesticides on ever-expanding monocultures; and the monopolization of seeds, which is to say, of the genetic resources on which all of humanity depends.[26]

The opponents of the measure argue that 'Proposition 37 is a deceptive, deeply flawed food labelling scheme, full of special-interest exemptions and loopholes.'[27] The opponents allege that the regime would 'create new government bureaucracy costing taxpayers millions, authorize expensive shakedown lawsuits against farmers and small businesses, and increase family grocery bills by hundreds of dollars per year'.[28]

Dr David Zilberman, a Professor of Agriculture from the University of California, Berkeley, was a champion for the No Campaign to Proposition 37.[29] He observed: 'Of course I am for people's right to know, but in the same way that you can label G-modified food, you can also label non-G-modified food.'[30] He noted: 'Today, if you don't really want G-modified food, you can buy organic, and there are

---

23 Ibid.
24 Ibid.
25 Ibid.
26 Ibid.
27 California General Election, *Proposition 37*, above n 4.
28 Ibid.
29 Democracy Now. *Food fight: Debating Prop 37, California's landmark initiative to label GMO food.* [Online]. Available at: http://www.democracynow.org/2012/10/24/food_fight_debating_prop_37_californias [accessed 2 April 2014].
30 Ibid.

[sic] also voluntary labeling of non-G-modified food' and contended that 'in every food system, you have some element of a mainstream food, things that are not being labelled.'[31] Zilberman observed that the proposition was costly, badly drafted, and based on the wrong promise and he denied accusations of corporate influence.[32] He maintained: 'Almost all the food that we eat is genetically modified.'[33] Zilberman makes arguments, both about the quality of the scheme, and its impact upon business.

Beyond the clear advocates and opponents there was some disquiet in the debate over the participation of food companies and biotechnology companies. The *Just Label It* movement complained:

> Transparency is our right. Yet a handful of companies, such as Monsanto, Kraft, Kellogg's and General Mills, have gotten away with hiding important information about our food for more than two decades ... For these companies, a defeat of labelling supports their interests, not the consumers. The Grocery Manufacturers Association (GMA) – Big Food's national lobby group – called defeating Prop 37 'the single highest priority for GMA' in 2012, and has already poured millions into defeating Washington State's initiative. While claiming their products are safe and that biotechnology is beneficial, they are emptying their pockets to fight a simple label.[34]

The *Just Label It* movement maintained: 'Polling shows overwhelming public support for labelling of genetically engineered foods, yet the same food and chemical companies continue to ignore consumers fight [for] our right to know every chance they get.'[35] The group lamented: 'This will continue to be a David versus Goliath battle, an unequal fight between the American consumer and corporate money'[36] and argued: 'It's time to call out these companies, and demand that [they] support the consumer's right to transparency.'[37]

Support for Proposition 37 faded as the Californian Election approached. In October 2012 a poll by the California Business Roundtable and the Pepperdine University School of Public Policy showed 39.1per cent of likely voters supported the measure, while 50.5 per cent opposed the labelling requirement.[38] The poll reported that undecided voters represented 10.5 per cent of respondents. At the

---

31   Ibid.

32   Ibid.

33   Ibid.

34   *Just Label It, Labeling opponents: Who are the companies fighting our right to know?* [Online]. Available at: http://justlabelit.org/right-to-know/labeling-opponents/ [accessed 2 April 2014].

35   Ibid.

36   Ibid.

37   Ibid.

38   Lifsher, M. 2012. Proposition 37 losing in late Business Roundtable-Pepperdine poll. *The Los Angeles Times*, 30 October 2012. [Online]. Available at: http://articles.latimes.com/2012/oct/30/business/la-fi-mo-37-losing-in-poll-20121030 [accessed 2 April 2014].

election on 6 November 2012 there were 6,088,714 votes in favour – 48.59 per cent of the votes cast – and 6,442,371 votes against – 51.45 per cent of the votes cast.[39]

Leon Kaye argued that the outcome highlighted the need to reform California's initiative process: 'Whatever your opinion is on Proposition 37 ... one issue is clear: California's ballot initiative process is in desperate need of reform.'[40] He observed: 'The collapse of support for Proposition 37 is a textbook case of how opponents of such a measure can find success by funding a negative campaign that confuses and jades voters.'[41] Kaye was concerned that the opponents of Proposition 37 relied heavily on out-of-state money: 'Californians' have got to find a way to limit the influence of companies whose operations are based outside of the state.'[42] He lamented: 'The decision should be made by debate and analysis of the facts; not a $45.6 million effort generated to buy an electoral outcome.'[43] In a fall op.-ed supporting GMO labels *New York Times* food writer Mark Bittman wrote, 'as goes California, so goes the nation.'[44]

In 2013 there was discussion of whether there will be further legislative initiatives in California. The Center for Food Safety commented:

> Disinformation won the day, but it did not change the facts about what California voters think of GE food labelling. This poll shows that the more the truth about Proposition 37 was received by voters, the more they voted for it. It's a certainty that once the money-induced cloud of doubt was lifted, many Californians viewed labelling of GE foods as the smart choice.[45]

The survey noted: 'While the state's hotly contested GE food labelling initiative was defeated by less than a 3 per cent margin, a full 67 per cent of voters continue to support the labelling of GE foods.'[46] The Center for Food Safety highlighted the

---

39   California Proposition 37, *Mandatory Labeling of Genetically Engineered Food.* [Online]. Available at: http://ballotpedia.org/California_Proposition_37,_Mandatory_ Labeling_of_Genetically_Engineered_Food_%282012%29 [accessed 2 April 2014].

40   Kaye, L. 2012. Near death of Proposition 37 proves California's initiative process needs reform. *Triple Pundit*, 6 November 2012. [Online]. Available at: http://www. triplepundit.com/2012/11/death-proposition-37-proves-californias-initiative-process-reform/ [accessed 2 April 2014].

41   Ibid.

42   Ibid.

43   Ibid.

44   Bittman, M. 2012. GMO's [sic]: Let's label'em. *The New York Times*, 15 September 2012. [Online]. Available at: http://opinionator.blogs.nytimes.com/2012/09/15/ g-m-o-s-lets-label-em/?_php=true&_type=blogs&_r=0 [accessed 2 April 2014].

45   Center for Food Safety. 2013. *'Post-Proposition 37 poll shows strong public support for future GE food labelling.* 9 January 2013. [Online]. Available at: http:// www.centerforfoodsafety.org/press-releases/781/post-prop-37-poll-shows-strong-public-support-for-future-ge-food-labeling [accessed 2 April 2014].

46   Ibid.

role of opponents of Proposition 37 – including Monsanto, DuPont, Dow, PepsiCo and Kraft.[47]

More recently, in 2014, Senator Noreen Evans has been agitating for law reform, putting forward new legislation requiring GMO labelling in California.[48] The Senate Bill 1381 is being pitched as a simpler version of the unsuccessful Proposition 37.[49]

State Battles over GM Food Labelling

In addition to the landmark proposition in California, there has been a range of initiatives in other states[50] including grassroots movements to improve the oversight of GM food.[51] It is worth highlighting the debates in Connecticut, Maine, Vermont, Washington, and Colorado. Such jurisdictions have been at the forefront of efforts for law reform in this field. A number of conservative states – like Texas – remain resistant to proposals in respect of GM food labelling.[52]

Connecticut

In June 2013 Connecticut was the first state to pass legislation on GM food labelling – with the caveat that the legislation would only come into effect when four other neighbouring states passed similar bills. Representative Republican John Shaban reflected upon the initiative:

> The House and Senate recently passed the 'GMO Bill' that will require the labelling of food containing genetically modified organisms (GMOs) once neighbouring states adopt similar provisions. GMOs are introduced into the genetic code of certain crops to promote particular characteristics such as a resistance to certain pesticides. The bill was prompted by the national debate regarding the potential

47   Ibid.

48   Moore, D. 2014. Evans' GMO food-labelling clears state Senate Committee. *The Press Democrat*, 26 March 2014. [Online]. Available at: http://www.pressdemocrat.com/article/20140326/articles/140329630 [accessed 2 April 2014].

49   Flynn, D. 2014. GE labelling resurrected in California, Petition for ballot measure circulating in Colorado. *Food Safety News*, 25 March 2014. [Online]. Available at: http://www.foodsafetynews.com/2014/03/gm-labeling-resurrected-in-california-petition-circulating-for-initiative-in-colorado/#.UzO4gxAXL-k [accessed 2 April 2014].

50   Center for Food Safety. 2013. *State labelling initiatives*. [Online]. Available at: http://www.centerforfoodsafety.org/issues/976/ge-food-labeling/state-labeling-initiatives# [accessed 2 April 2014].

51   See Right to Know GMO. *Map*. [Online]. Available at: http://www.righttoknow-gmo.org/states [accessed 2 April 2014].

52   Ahmed, A. 2014. The push to label genetically modified products. *The New York Times*, 22 March 2014. [Online]. Available at: http://www.nytimes.com/2014/03/23/us/the-push-to-label-genetically-modified-products.html [accessed 2 April 2014].

health effects of ingesting GMOs ... The bill makes Connecticut the leader on this effort, and should create the spark needed to effect a regional or national labelling model driven by both government and market participants.[53]

The Governor, Dannel Malloy, agreed that he would sign the bill into law – after reaching an agreement with the legislature that the law would not take effect unless four other states passed similar regulations.[54] He commented: 'This bill strikes an important balance by ensuring the consumers' right to know what is in their food while shielding our small businesses from liability that could leave them at a competitive disadvantage.'[55] The neighbouring states clause seems to be a compromise to ensure that Connecticut will not be punished or adversely affected for being a first mover.

Maine

In February 2013, the State of Maine legislature introduced *An Act to Protect Maine Food Consumers' Right to Know about Genetically Engineered Food and Seed Stock*. According to the Bill's summary, the legislation:

> requires disclosure of genetic engineering at the point of retail sale of food and seed stock and provides that food or seed stock for which the disclosure is not made is considered to be misbranded and subject to the sanctions for misbranding. The bill provides that food or seed stock may not be labelled as natural if it has been genetically engineered. The bill exempts products produced without knowledge that the products, or items used in their production, were genetically engineered; animal products derived from an animal that was not genetically engineered but was fed genetically engineered food; and products with only a minimum content produced by genetic engineering. The bill also provides that the disclosure requirements do not apply to restaurants, alcoholic beverages or medical food. The disclosure provisions are administered by the Department of Agriculture, Conservation and Forestry.[56]

On the 11 June 2013 the Maine House passed the legislation by a vote of 141 to 4. The Maine Senate unanimously supported the labelling of genetically modified

---

53   Shaban, J. 2013. *Passage of GMO labelling legislation.* 5 June 2013. [Online]. Available at: https://cthousegop.com/2013/06/rep-shaban-on-passage-of-gmo-labeling-legislation/ [accessed 2 April 2014].

54   Stephanie Strom, Connecticut approves labelling genetically modified foods. *The New York Times*, 3 June 2013. [Online]. Available at: http://www.nytimes.com/2013/06/04/business/connecticut-approves-qualified-genetic-labeling.html?adxnnl=1&adxnnlx=13964 00602–757tRA0JESXtaUE+NOmcug [accessed 2 April 2014].

55   Ibid.

56   The Right to Know GMO. *Maine*. [Online]. Available at: http://righttoknow-gmo. org/states/maine [accessed 2 April 2014].

foods. On 9 July 2013, Governor LePage pledged to sign the bill making it law in the state of Maine. In a letter sent to Representative Lance Harvell (Republican-Farmington) and Senator Chris Johnson (Democrat-Lincoln County), lead sponsors of the bill, Governor LePage stated: 'I deeply appreciate the strong public sentiment behind the bill and agree that consumers should have the right to know what is in their food.'[57] The Governor noted: 'Additionally, my support for the bill is based in large part on the requirement in the bill that similar legislation be enacted and passed in other contiguous states.'[58]

Legislator Sharon Treat from the State of Maine has been concerned about the regime being affected by international trade agreements.[59] She commented upon the desire of Maine to protect its unique local laws:

> In our state of Maine, which is a rather low-income state with limited economic opportunity (especially now that our textile and shoe factories have almost all moved offshore following NAFTA and other trade agreements), a bright spot is local food initiatives. Our land use and procurement policies are encouraging young people to take up farming, and developing new markets for farmers to sell their produce to schools, hospitals, and other institutions. We have enacted a GMO labeling law similar to that in effect in EU countries, and policies that encourage organic and niche farming. We have also enacted procurement laws – in effect for over a decade – which do not permit the purchase by our state government of products made pursuant to unfair labor practices, or where discrimination is permitted.[60]

Sharon Treat was concerned that Maine's food labelling regulations and environmental laws would be under threat from trade deals, such as the *Trans-Pacific Partnership* and the *Trans-Atlantic Trade and Investment Partnership*.

Vermont

In 2005 Vermont considered the adoption of the *Farmer Protection Act* to deal with GM crops.[61] The bill placed strict liability for any economic damage

57   Maine Organic Farmers and Gardeners Association. *Governor LePage pledges support for GMO labelling.* [Online]. Available at: http://www.mofga.org/Programs/PublicPolicyInitiatives/RightToKnowGMOMaine/LePagePledge/tabid/2645/Default.aspx [accessed 2 April 2014].

58   Ibid.

59   Treat, S. 2014. *Concerns about TTIP not just in Europe: Interview with US state legislator, TTIP: Beware what lies beneath, 26 March 2014.* [Online]. Available at: http://ttip2014.eu/blog-detail/blog/USA%20concerns.html [accessed 2 April 2014].

60   Ibid.

61   Carrick, D. 2005. GM contamination of crops, *The law report, ABC.* [Online]. Available at: http://www.abc.net.au/radionational/programs/lawreport/gm-contamination-of-crops/3373262 [accessed 2 April 2014].

to farmers on the producers of GM products, the biotech firms. Under the proposal, when farmers sustain any damages as a result of contamination, they can claim compensation without having to prove the company's negligence. Joe Mendelson from the Center for Food Safety explained the impetus for the measure.[62] There was a great debate in Vermont about that particular piece of legislation. Some members of the Vermont General Assembly believe that such legislation was unnecessary. One member said that it was like having a baseball bat to attack a gnat.[63] However, others believed that such legislation was very important in protecting farmers and growers from liability concerns in relation to GM crops.

In 2011 Representative Kate Webb of Shelburne was the lead sponsor of a bill requiring mandatory labels, the *VT Right to Know Genetically Engineered Food Act* 2011.[64] Although the bill was approved by the House Agriculture Committee by a vote of nine to one, the bill was not passed by the end of the legislative session. Interestingly, the biotechnology company Monsanto threatened to sue the state of Vermont if the legislative bill was passed.[65] In 2012 Corin Hirsch wondered: 'So with Proposition 37 dead in California, will Vermont become the first state to require labelling of GMOS?'[66] In 2013, supporters of the VT Right to Know GMO Coalition – including the Vermont Public Interest Research Group, the Northeast Farming Association of Vermont and Rural Vermont – sought to reintroduce a similar version of the bill.[67] Falko Schilling, a consumer advocate, commented: 'I think Vermont has a great opportunity to lead on this issue.'[68]

In May 2013, a GMO Labelling bill passed the Vermont House of Representatives.[69] However, some legislators opposed the bill. Governor Peter Shumlin was on record as being dubious about whether the bill would withstand a constitutional challenge. And the biotechnology industry has threatened to

---

62   Ibid.

63   Ibid.

64   Vermont Right to Know GMOs. *H.722, The VT Right to Know Genetically Engineered Food Act introduced to Vermont House of Representatives*. [Online]. Available at: http://www.vpirg.org/news/h-722-the-vt-right-to-know-genetically-engineered-food-act -introduced-to-vermont-house-of-representatives-2/ [accessed 2 April 2014].

65   Russia Today Question More. 2014. *Monsanto threatens to sue the entire state of Vermont*, 6 April 2014. [Online]. Available at: http://rt.com/usa/news/monsanto-sue-gmo-vermont-478/ [accessed 2 April 2014].

66   Hirsch, C. 2012. *With Prop 37 dead in California, will Vermont become the first to label GMOs?* 8 November 2012. [Online]. Available at: http://www.sevendaysvt.com/ BiteClub/archives/2012/11/08/with-prop-37-dead-in-california-will-vermont-become-the-first-to-label-gmos [accessed 2 April 2014].

67   Ibid.

68   Ibid.

69   *An Act relating to the Labeling of Food Produced with Genetic Engineering* 2012 (Vermont) H. 722. [Online]. Available at: http://www.leg.state.vt.us/docs/2012/bills/Intro/ H-722.pdf [accessed 2 April 2014].

challenge the validity and the legitimacy of any state laws in respect of GM food labelling. Monsanto has maintained:

> We oppose current initiatives to mandate labelling of ingredients developed from GM seeds in the absence of any demonstrated risks. Such mandatory labelling could imply that food products containing these ingredients are somehow inferior to their conventional or organic counterparts.[70]

Monsanto has threatened to challenge any Vermont legislation on GM food labelling.[71] In particular, it has placed reliance on the 1996 decision of the US Court of Appeals for the Second Circuit in *International Dairy Foods Association v. Jeffrey Amestoy, Attorney General of Vermont* on bovine somatotrophin (BST) labelling.[72] In this matter, dairy manufacturers challenged Vermont laws, which required dairy manufacturers to identify products that were derived from cows treated with a synthetic growth hormone used to increase milk production. The dairy manufacturers alleged that the legislation violated the Commerce Clause and the First Amendment of the *US Constitution.*

At first instance, the judge focused on the economic impact of labelling and found that the dairy manufacturers had not demonstrated irreparable harm to any right protected by the First Amendment.

On appeal, the court found in favour of the dairy manufacturers by a majority of two to one. For the majority, Judge Altimari held that the dairy manufacturers were entitled to an injunction. The judge held: 'Because the statute at hand unquestionably implicates the dairy manufacturers' speech rights, we reject the district court's conclusion that the disclosure compelled by Vt. Stat. Ann. tit. 6, § 2754(c), is not a "loss of First Amendment freedoms", amounting to irreparable harm.'[73] Altimari maintained: 'We do not doubt that Vermont's asserted interest, the demand of its citizenry for such information, is genuine; reluctantly, however, we conclude that it is inadequate.'[74] However, the judge observed: 'We are aware of no case in which consumer interest alone was sufficient to justify requiring a product's manufacturers to publish the functional equivalent of a warning about a production method that has no discernible impact on a final product'.[75] Judge Altimari held:

---

70   Knowles, D. Vermont Senator continues fight for GMO labelling, defeat of Monsanto Protection Act. *New York Daily News*, 28 May 2013. [Online]. Available at: http://www.nydailynews.com/news/politics/bernie-sanders-pressure-monsanto-article-1.1357031#ixzz2hf2EzMSK [accessed 2 April 2014].

71   Ibid.

72   *International Dairy Foods Association* v. *Attorney General of Vermont* 92 F.3d 67 (1996).

73   Ibid., 72.

74   Ibid., 73.

75   Ibid., 73.

Absent, however, some indication that this information bears on a reasonable concern for human health or safety or some other sufficiently substantial governmental concern, the manufacturers cannot be compelled to disclose it. Instead, those consumers interested in such information should exercise the power of their purses by buying products from manufacturers who voluntarily reveal it.[76]

The judge then considered that 'consumer curiosity alone is not a strong enough state interest to sustain the compulsion of even an accurate, factual statement.'[77]

In dissent, Judge Leval held that the First Amendment should support the use of labelling for the purposes of consumer protection and public health:

The policy of the First Amendment, in its application to commercial speech, is to favor the flow of accurate, relevant information. The majority's invocation of the First Amendment to invalidate a state law requiring disclosure of information consumers reasonably desire stands the Amendment on its ear. In my view, the district court correctly found that plaintiffs were unlikely to succeed in proving Vermont's law unconstitutional.[78]

Judge Leval commented that the true objective of the milk producers is concealment. The judge maintained: 'The question is simply whether the First Amendment prohibits government from requiring disclosure of truthful relevant information to consumers.'[79] His Honour concluded: 'In my view, the interest of the milk producers has little entitlement to protection under the First Amendment.'[80] Judge Leval stressed that 'the case law that has developed under the doctrine of commercial speech has repeatedly emphasized that the primary function of the First Amendment in its application to commercial speech is to advance truthful disclosure – the very interest that the milk producers seek to undermine.'[81] His Honour emphasised that, in any case, 'the precedential effect of the majority's ruling is quite limited' because 'it applies only to cases where a state disclosure requirement is supported by no interest other than the gratification of consumer curiosity.'[82]

The decision in this matter could be contrasted with the recent decision of the High Court of Australia in respect of the plain packaging of tobacco products – which held that health warnings were commonplace and that the Australian Government had the legislative power to enforce their

---

76 Ibid., 74.
77 Ibid., 74.
78 Ibid., 74.
79 Ibid., 75.
80 Ibid., 80.
81 Ibid., 81.
82 Ibid., 81.

use.[83] Justice Kiefel commented: 'Many kinds of products have been subjected to regulation in order to prevent or reduce the likelihood of harm. The labelling required for medicines and poisonous substances comes immediately to mind. Labelling is also required for certain foods, to both protect and promote public health.'[84] It should be noted that, in this particular case, the combination of graphic pictures and plain packaging of tobacco products were considered to be consistent with the *Australian Constitution*.

In 2014, the Governor of Vermont signed the GM food labelling bill. Governor Peter Shumlin discussed the legislative measure:

> Vermonters take our food and how it is produced seriously, and we believe we have a right to know what's in the food we buy. I am proud that we're leading the way in the United States to require labeling of genetically engineered food. More than 60 countries have already restricted or labeled these foods, and now one state – Vermont – will also ensure that we know what's in the food we buy and serve our families.[85]

Shumlin commented: 'There is no doubt that there are those who will work to derail this common sense legislation.'[86] He noted: 'As you know, we're in the middle of an agricultural renaissance in Vermont because more and more Vermonters care about where their food comes from, what's in it, and who grew it.'[87] Shumlin observed that the legislation would create momentum for change elsewhere in the US: 'It makes sense that we are again leading the nation in this important step forward.'[88]

The legislative bill created a special fund to support the implementation and administration of the state labelling law, including costs and fees associated with any legal challenge to the regime. The Vermont Attorney, General William Sorrell, observed: 'The constitutionality of the GMO labelling law will

---

83   *JT International SA v. Commonwealth of Australia* [2012] HCA 43 (5 October 2012). See Rimmer, M. 2014. The High Court of Australia and the Marlboro man: The battle over the plain packaging of tobacco products, in *Regulating Tobacco, Alcohol and Unhealthy Foods: The Legal Issues* edited by T. Voon, A. Mitchell and J. Liberman. London: Routledge, 337–60; Rimmer, M. 2013. Plain packaging for the Pacific Rim: The Trans-Pacific Partnership and tobacco control, in *Trade Liberalisation and International Co-operation: A Legal Analysis of the Trans-Pacific Partnership Agreement* edited by T. Voon. Cheltenham (UK) and Northampton (Mass.): Edward Elgar, 75–105.

84   *JT International SA v. Commonwealth of Australia* [2012] HCA 43 (5 October 2012) [316].

85   Shumlin, P. 2014. *Governor signs first in the nation genetically engineered foods labeling law*, 8 May 2014. [Online]. Available at: http://governor.vermont.gov/newsroom-gmo-bill-signing-release [accessed 2 April 2014].

86   Ibid.

87   Ibid.

88   Ibid.

undoubtedly be challenged'. He commented: 'I can promise that my office will mount a vigorous and zealous defense of the law that has so much support from Vermont consumers.'[89]

It is anticipated that Vermont's GM food labelling laws will be challenged under a number of grounds by its opponents – including in respect of the First Amendment, Federal pre-emption, and constitutional laws regarding commerce and acquisition of property.[90]

The Independent Senator from Vermont in the US Congress, Bernie Sanders, was supportive of Vermont's bill.[91]

Washington State

In January 2013 the bill – SB. 5073 was introduced to provide for the labelling of genetically engineered foods and prescribe penalties for violation. The bill was sponsored by Senators Chase, Klein, Keiser, Rolfes and Hasegawa. Although this bill was not passed, it nonetheless provided the impetus for public debate and led to a citizen initiated initiative.

The State of Washington had a citizen initiated initiative relating to I-522.[92] This initiative is entitled 'an act relating to disclosure of foods produced through genetic engineering'. Section One served as a preamble or a recital. The bill maintained that 'polls consistently show that the vast majority of the public, typically more than ninety per cent, wants to know if their food was produced using genetic engineering.'[93] The bill warned: 'without disclosure, consumers of genetically engineered food unknowingly may violate their own dietary and religious restrictions.'[94] The bill observed that there was a gap in the legal framework in the US:

> Currently, there is no federal or state law that requires food producers to identify whether foods were produced using genetic engineering. At the same time, the United States Food and Drug Administration does not require safety studies of such foods. Unless these foods contain a known allergen, the United States food and drug administration does not require the developers of genetically engineered crops to consult with the agency. Consultations with the United

---

89   Ibid.

90   Chokshi, N. 2014. 'Vermont Just Passed The Nation's First GMO Food Labeling Law. Now it Prepares to Get Sued,' *The Washington Post*, 9 May 2014. [Online]. Available at: http://www.washingtonpost.com/blogs/govbeat/wp/2014/04/29/how-vermont-plans-to-defend-the-nations-first-gmo-law/ [accessed 2 April 2014].

91   Knowles, D. 2013. above n 70.

92   *Yes on 522*. [Online]. Available at: http://yeson522.com/about/read/ [site now archived].

93   Ibid.

94   Ibid.

States food and drug administration are entirely voluntary and the developers themselves may decide what information they may wish to provide.[95]

The bill contended that 'mandatory identification of foods produced with genetic engineering can provide a critical method for tracking the potential health effects of consuming foods produced through genetic engineering.'[96] Moreover, the sponsors of the bill maintained: 'Consumers have the right to know whether the foods they purchase were produced with genetic engineering.'[97] The bill observed: 'Forty-nine countries, including Japan, South Korea, China, Australia, New Zealand, Thailand, Russia, the European Union member states and other key US trading partners, have laws mandating disclosure of genetically engineered foods on food labels.'[98] The bill stressed: 'Many countries have restrictions or bans against foods produced with genetic engineering,'[99] Finally, the bill claimed: 'No international agreements prohibit the mandatory identification of foods produced through genetic engineering.'[100]

Section 2 of the bill provides definitions of key words in the legislative proposal. Section 3 provides the key prescriptions in respect of GM food labelling:

Beginning July 1, 2015, any food offered for retail sale in Washington is misbranded if it is, or may have been, entirely or partly produced with genetic engineering and that fact is not disclosed as follows:

(a) In the case of a raw agricultural commodity, on the package offered for retail sale, with the words 'genetically engineered' stated clearly and conspicuously on the front of the package of such a commodity, or in the case of such a commodity that is not separately packaged or labelled, on a label appearing on the retail store shelf or bin where such a commodity is displayed for sale;

(b) In the case of any processed food, on the front of the package of such food produced by a manufacturer, with the words 'partially produced with genetic engineering' or 'may be partially produced with genetic engineering' stated clearly and conspicuously; and

(c) In the case of any seed or seed stock, on the seed or seed stock container, sales receipt or any other reference to identification, ownership, or possession, with the words 'genetically engineered' or 'produced with genetic engineering' stated clearly and conspicuously.[101]

---

95   Ibid.
96   Ibid.
97   Ibid.
98   Ibid.
99   Ibid.
100   Ibid.
101   Ibid.

However, the legislation does 'not require either the listing or identification of any ingredient or ingredients that were genetically engineered, nor that the term "genetically engineered" be placed immediately preceding any common name or primary product descriptor of a food'.[102] The legislation contains a number of exemptions from the operation of this scheme.

Section 4 provides that 'the department may adopt rules necessary to implement this chapter, provided that the department is not authorized to create any exemptions beyond those provided in section 3(3) of this act.'

Section 5 provides for penalties. Section 5 (2) noted: 'The department may assess a civil penalty against any person violating this chapter in an amount not to exceed one thousand dollars per day. Section 5 (3) provides 'An action to enjoin a violation of this chapter may be brought in any court of competent jurisdiction by any person in the public interest if the action is commenced more than sixty days after the person has given notice of the alleged violation to the department, the attorney general, and to the alleged violator.' Section 5 (4) stipulates: 'The court may award to a prevailing plaintiff reasonable costs and attorneys' fees incurred in investigating and prosecuting an action to enforce this chapter.'[103]

Carole Bartolotto – a dietician – supported the labelling of GMOs in the state of Washington: 'Washington's Initiative 522 to label genetically engineered foods, on the November ballot, will help us get the transparency we desire.'[104] She maintained that 'If Washington's Initiative 522 passes and genetically modified foods are labelled … it just might change the face of American agriculture forever.'[105]

There has been concern that the debate over GM food labelling in Washington has been shaped by industry contributions.[106] A journalist reported that opponents of the initiative had raised $17.2 million, while its supporters had only raised $4.9 million:

> The money raised so far by both sides, about $21.9 million, is the second highest amount for a state ballot measure, according to records kept by the Washington Public Disclosure Commission. It trails money raised for and against a 2011 measure to privatize liquor sales.
>
> Nearly all of the opposition money against I-522 has come from six out-of-state contributors that also were among the top donors against California's

---

102   Ibid.

103   Ibid.

104   Bartolotto, C. 2013. Why genetically modified food should be labelled. *The Huffington Post*, 4 October 2013. [Online]. Available at: http://www.huffingtonpost.com/carole-bartolotto/why-genetically-modified-food_b_4039114.html [accessed 2 April 2014].

105   Ibid.

106   Le, P. Big money shapes GMO food labelling fight in Washington. *Associated Press*, 6 October 2013. [Online]. Available at: http://www.komonews.com/news/local/Big-money-shapes-GMO-food-labeling-fight-in-Wash--226675431.html [accessed 2 April 2014].

measure. The Grocery Manufacturers Association has given $7.2 million, while the biotechnology company Monsanto Co. has given $4.8 million. The average contribution to No on 522 is $1.2 million.

About 72 per cent of the money raised by supporters of I-522 has also come from out of state. Dr. Bronner's Magic Soaps has given the most: $1.7 million. The pro-labelling group has received lots of smaller donations from within the state. The average contribution to Yes on 522 is $874.[107]

Environmentalists have urged food industry and biotechnology industry groups to desist from funding the campaign against the proposed food labelling law in Washington State. Lucia von Reusner of Green Century Capital Management, manager of environmentally focused mutual funds, argued: 'We believe that political contributions are a poor investment and are calling companies not to spend money opposing legislation that would give consumers labelling information.'[108]

In November 2013 Washington State voters rejected the initiative. The vote was 45.2 per cent in favour of labelling, and 54.8 per cent opposed to labelling.[109] Marion Nestle, Professor of Nutrition at New York University, predicted that there would be further regulatory efforts: 'At some point the industry is going to get tired of pouring this kind of money into these campaigns.'[110] No doubt there will be further efforts to push for GM food labelling in Washington State.

Colorado

In 2014 there was an effort to put forward a ballot initiative.[111] Initiative #48 would require that GM food came with packaging that announced 'produced with genetic engineering' by 1 July 2016. This initiative was challenged in the Supreme Court of Colorado.[112] The Rocky Mountain Food Industry Association President, Mary Lou Chapman, and Mark Arnusch, a farmer in rural Keensburg, brought a challenge alleging that the ballot was misleading. The Supreme Court of Colorado

---

107   Ibid.

108   Abbott, C. 2013. Greens ask US biotech firms to sit out food-labelling vote. *Reuters*, 9 October 2013. [Online]. Available at: http://www.reuters.com/article/2013/10/10/usa-agriculture-gmo-idUSL1N0HZ1OJ20131010 [accessed 2 April 2014].

109   Weise, E. 2013. Washington State voters reject state labeling of GMO foods. *USA Today*, 6 November 2013. [Online]. Available at: http://www.usatoday.com/story/news/nation/2013/11/06/washington-state-voters-reject-gmo-labeing/3450705/ [accessed 2 April 2014].

110   Ibid.

111   RT Question More. 2014. *GMO labelling effort in Colorado scores win in State Supreme Court*, 19 March 2014. [Online]. Available at: http://rt.com/usa/colorado-gmo-label-court-945/ [accessed 2 April 2014].

112   *Arnusch and Chapman* v. *Cooper and Gray* (2014) Colorado Supreme Court, Case Number L 2013SA335.

dismissed action. Larry Cooper, one of the individuals who proposed the initiative, observed: 'We are pleased that the state Supreme Court ruled in favor of the GMO labelling ballot title, and we look forward to bringing a GMO labelling initiative before the voters of Colorado this fall.'[113] He stressed: 'Coloradans have the right to know what is in their food, and to make purchasing decisions for their families based on knowing whether their foods are genetically engineered, and we believe they will have that opportunity after November.'[114]

It remains to be seen whether there will be a larger coalition of states, proposing regimes dealing with GM food labelling, given the closely contested nature of the initiatives.[115]

President Barack Obama, the US Congress and GM Food Labelling

Turning from State to Federal legislation, in 2007, as a Presidential candidate, Barack Obama, supported GM food labelling, emphasising that he would strive to 'let folks know when their food is genetically modified, because Americans have a right to know what they are buying'.[116] However, he has shown little enthusiasm for policy reform during his two terms of Presidency. In his piece, 'The Vote for the Dinner Party', Michael Pollan observed that the Food Movement needed to gain greater influence in Federal Politics:

> That's why, sooner or later, the food movement will have to engage in the hard politics of Washington – of voting with votes, not just forks. This is an arena in which it has thus far been much less successful. It has won little more than crumbs in the most recent battle over the farm bill (which every five years sets federal policy for agriculture and nutrition programs), a few improvements in school lunch and food safety and the symbol of an organic garden at the White House. The modesty of these achievements shouldn't surprise us: the food movement is young and does not yet have its Sierra Club or National Rifle Association, large membership organizations with the clout to reward and punish legislators. Thus while Big Food may live in fear of its restive consumers, its grip on Washington has not been challenged.[117]

---

113 Long, J. Colorado Supreme Court OKs ballot title on GMO labelling initiative. *Natural Products Insider*, 19 March 2014. [Online]. Available at: http://www.naturalproductsinsider.com/news/2014/03/colorado-supreme-court-oks-ballot-title-on-gmo-la.aspx [accessed 2 April 2014].

114 Ibid.

115 Flynn, D. 2014. above n 49.

116 Philpott, T. 2011. Obama's broken promise on GMO food labelling. *Mother Jones*, 6 October 2011. [Online]. Available at: http://www.motherjones.com/tom-philpott/2011/10/fda-labeling-gmo-genetically-modified-foods [accessed 2 April 2014].

117 Pollan, M. 2012, above n 22.

The key features of Federal involvement with GM food labelling have been the indecisive debate about the role of the FDA in respect of the regulation of GM food labelling, concerted efforts in 2013 to introduce legislative measures in the US Congress and the response by Congressional supporters of biotechnology to counter GM food labelling efforts at federal and state levels.

The US Food and Drug Administration

In 1992 the FDA published a policy statement on foods created through genetic engineering.[118] The policy allowed for genetically engineered foods to be marketed without labelling. The policy was based upon the determination that GM foods were substantially equivalent to foods produced through conventional methods. There has been much debate and controversy over this policy statement. In 2011 the Center for Food Safety petitioned the FDA to reform its regulation.[119] The Center maintained:

> Genetic engineering results in changes to foods at the molecular level that have never occurred in traditional varieties. These changes are determinative of consumers' food purchases and not readily apparent. Thus, the absence of mandatory labelling disclosures for GE foods is misleading to consumers. FDA's failure to require labelling for GE foods is an abdication of its statutory mandate to require labelling for foods that are "misbranded" because they are misleading.[120]

The petitioners demanded that the FDA require 'that foods that are genetically engineered organisms, or contain ingredients derived from genetically engineered organisms – collectively referred to as "GE foods" – be labelled under the *Federal Food Drug and Cosmetic Act*'.[121] The Center contended that 'the requested actions are necessary to prevent economic fraud, and to protect consumers who are deceived by thinking the absence of labelling means the absence of GE foods.'[122]

The Center lamented that the current regulatory regime was inadequate and insufficient for GM food labelling:

---

118   FDA, 1992, Statement of Policy – Foods Derived from New Plant Varieties, *Federal Register*, Vol. 57 (104), 22984. [Online]. Available at: http://www.fda.gov/food/guidanceregulation/guidancedocumentsregulatoryinformation/biotechnology/ucm096095.htm [accessed 2 April 2014].

119   *Center for Food Safety v. United States Food and Drug Administration* 2011. [Online]. Available at: http://www.centerforfoodsafety.org/files/ge-labeling-petition-10–11–2011-final1_21309.pdf [accessed 2 April 2014].

120   Ibid., 2.

121   Ibid.

122   Ibid.

FDA's outdated regulatory regime for food labelling is woefully inadequate. FDA is still using 19th century ideas to regulate 21st century foods, focusing only on traits that consumers can detect with their senses. But modern public preferences and purchasing decisions are based not only on sensory perceptions, but also on concerns related to latent or unknown health risks, animal welfare, faith, political concerns, social justice, and environmental impacts. In addition to genetic engineering, other novel and unnatural food production technologies are either on the horizon or are currently in use, many completely unbeknownst to consumers.[123]

The Center argued 'The use of these novel food technologies on a commercial scale has so far slipped underneath FDA's current threshold for 'materiality' because they make silent, genetic, and molecular changes to food that are not capable of being detected by human senses.'[124] The Center was concerned about the impact of such policies upon consumers: 'As the use of these and future food production technologies proliferates, consumers know less and less about the food they put in their bodies.'[125]

The Center stressed that 'the power and duty to modernize the oversight of food lies with Food and Drug Administration.'[126] The Center argued that the 'failure to require labeling of GE foods conflicts with this past FDA precedent and creates the appearance that FDA has altered its past policies to benefit the biotechnology industry, not the public'.[127] The Center submitted that the 'FDA has not just the statutory authority, but also the duty to require that products of novel food technologies, particularly genetic engineering, be labeled differently from their conventional counterparts.'[128] The Center maintained that the 'FDA's failure to take the requested action would be arbitrary, capricious, and contrary to law'.[129] The Center concluded:

> Genetic engineering makes silent but fundamental changes to our food at the molecular and cellular level, the full human health and environmental consequences of which are still being discovered. Unlabelled GE foods are misleading to consumers, who in the absence of labelling overwhelmingly purchase based on the reasonable assumption that their food is produced conventionally. Mandatory labelling for GE foods is necessary in order to prevent consumer deception and economic fraud.[130]

---

123 Ibid., 7.
124 Ibid.
125 Ibid.
126 Ibid., 8.
127 Ibid.
128 Ibid.
129 Ibid.
130 Ibid., 20.

The FDA, though, has been unyielding in the face of such entreaties for the introduction of GM food labelling. The Administration entirely ignored the prominent issue of GM food labelling in new food labelling rules in 2014. Ronnie Cummins of the Organic Consumers Association welcomed new regulations on nutrition: 'Changes to nutrition labels are long overdue, and it's great that Mrs Obama is leading the charge to force food manufacturers to provide more accurate information about their products.'[131] However, he lamented that 'conspicuously absent from the media hype was any mention of the one label that consumers have been crystal clear about wanting, the label that consumers in nearly 60 other countries have but Americans don't – a label that tells us whether or not our cereal or soda or mac and cheese contains genetically modified organisms (GMOs).'[132]

The Genetically Engineered Food Right to Know Act 2013 (US)

In April 2013, US Senator Barbara Boxer, a Democrat for California, and Congressman Peter DeFazio, a Democrat from Oregon, introduced the *Genetically Engineered Food Right-to-Know Act* 2013 (US).[133] The legislation would require the FDA to mandate clearly labelled genetically engineered foods. Senator Boxer commented on the legislative bill:

> *Americans have the right to know what is in the food they eat so they can make the best choices for their families. This legislation is supported by a broad coalition of consumer groups, businesses, farmers, fishermen and parents who all agree that consumers deserve more – not less – information about the food they buy.*[134]

Congressman DeFazio commented: 'When American families purchase food, they deserve to know if that food was genetically engineered in a laboratory.'[135] He observed: 'This legislation is supported by consumer's rights advocates, family farms, environmental organizations, and businesses, and it allows consumers to make an informed choice.'[136] Senator Gillibrand maintained: 'American consumers have made it clear that they want to be empowered to make choices about the food they eat.'[137] The Senator stressed: 'This legislation will deliver

---

131   Cummins, R. 2014. New FDA food label rules ignore the GMO elephant in the room. *The Huffington Post*, 25 March 2014. [Online]. Available at: http://www.huffingtonpost.com/ronnie-cummins/new-fda-food-label-rules-_b_5022900.html [accessed 2 April 2014].

132   Ibid.

133   Boxer, B. 2013. Boxer, DeFazio introduce Bill to require labelling of genetically engineered foods. *Press Release*, 24 April 2013. [Online]. Available at: http://www.boxer.senate.gov/en/press/releases/042413.cfm [accessed 2 April 2014].

134   Ibid.

135   Ibid.

136   Ibid.

137   Ibid.

the transparency every American deserves by providing clear labelling standards for food containing genetically engineered ingredients.'[138] Senator Richard Blumenthal commented on the legislative proposal:

> This is a common sense approach to ensuring that American consumers know more and make more informed decisions about the foods they eat. As an advocate for consumers' rights and ally of many groups supporting this measure, I want to make sure the food industry gives consumers the full story about what they put on their dinner tables. Consumers deserve to have clear, consistent, and accurate facts about the food products they purchase. More information is always better than less.[139]

Alaskan Senator Begich maintained: 'Labelling Genetically Engineered food should be a no-brainer which is why I'm pleased to join my colleagues on this bill to make sure consumers are fully informed when they make choices at the grocery store.'[140] Senator Tester said: 'American families shouldn't have to play a guessing game when it comes to the food they put on their kitchen tables.' The Senator insisted that 'consumers have a right to know what's in their food, and this bill gives them the tools they need to make informed decisions about the foods they choose.'[141] Vermont Independent Senator Bernie Sanders observed: 'All over this country people are becoming more conscious about the foods they are eating and the foods they are serving to their kids.'[142]

Oregon Senator Merkley said: 'Labelling is the common sense way to bring more transparency to consumers.'[143] Congressman Jared Polis, a Democrat from Colorado, supported the measure in terms of consumer rights.[144] He commented:

> Despite the prevalence of Genetically Modified Organisms (GMOs) in grocery stores and prepared foods, it remains difficult if not impossible for consumers to determine if the foods they eat contain GMOs. This labelling bill is about empowering consumers: consumers can choose to eat or not eat GMOs, or to pay more or less for GMOs. I believe consumers have a right to know what they are eating so they can make their own informed food choices. I am proud to be working toward more informative food labels.[145]

---

138 Ibid.
139 Ibid.
140 Ibid.
141 Ibid.
142 Ibid.
143 Ibid.
144 Polis, J. 2013. *Polis, Defazio introduce Bill to require labelling of genetically engineered food.* 24 April 2013. [Online]. Available at: http://polis.house.gov/news/docu mentsingle.aspx?DocumentID=331458 [accessed 2 April 2014].
145 Ibid.

The press release noted: 'Unfortunately, the FDA's antiquated labelling policy has not kept pace with 21st century food technologies that allow for a wide array of genetic and molecular changes to food that can't be detected by human senses.'[146] The press release invoked common sense: 'Common sense would indicate that GE corn that produces its own insecticide – or is engineered to survive being doused by herbicides – is materially different from traditional corn that does not.'[147] The press release also made comparisons to patent law: 'Even the US Patent and Trademark Office has recognized that these foods are materially different and novel for patent purposes.'[148]

In an interview with *The Huffington Post*, De Fazio – who has grown organic produce – said that he was agnostic about the health effects of GMOs.[149] However, he supported the mandatory labelling of food with genetically-engineered ingredients because of a strong conviction about the importance of consumer choice: 'Even the most ardent free market advocate, someone who's a devout follower of Adam Smith, would have to admit that consumers aren't being given full information right now.'[150]

De Fazio was hopeful that the legislative bill would generate a 'grassroots tidal wave of support' from voters – much as the National Organic Standards did in 1993. Nonetheless, he was concerned that the Obama administration was rather indifferent to the proposal: 'They're approaching it more like a competitive biotech issue for the US, as opposed to a much more insidious threat to our farmers and to consumers.'[151]

Scott Faber, president of the Environmental Working Group and the *Just Label It* campaign in favour of GMO labelling, expected opposition from the biotechnology and agricultural industries and predicted that the bill 'faces an uphill climb in both the House and Senate'.[152] Biotechnology Industry Organization (BIO) spokeswoman, Karen Badt, signalled that the biotechnology industry would oppose the proposal in relation to food labelling: 'Unfortunately, advocates of mandatory "GMO labelling" are working an agenda to vilify biotechnology and scare consumers away from safe and healthful food products.'[153]

Elizabeth Kucinich, the Policy Director of the Center for Food Safety, has called for President Barack Obama to fulfil his campaign promises on GMO

146　Ibid.
147　Ibid.
148　Ibid.
149　Satran, J. 2013. Genetically engineered food labelling taken on by Congress in Right-to-Know Act. *The Huffington Post*, 25 April 2013. [Online]. Available at: http://www.huffingtonpost.com/2013/04/25/genetically-engineered-food_n_3149418.html?utm_hp_ref=tw [accessed 2 April 2014].
150　Ibid.
151　Ibid.
152　Ibid.
153　Ibid.

labelling.[154] She stressed that 'Although the FDA doesn't need congressional authorization in order to mandate a federal labeling standard for GMOs, federal lawmakers have become increasingly vocal on the issue.'[155] Kucinich observed:

> There is ample precedent. As well as basic ingredients, labels also disclose country of origin, irradiation and even orange juice 'made from concentrate'. FDA's labelling requirements are not based solely on safety concerns and nutrition, a common myth propped up by the food and chemical industries. A federal labelling standard would give consumers the opportunity to make their own choices about the foods they bring home to their families.[156]

In her view, President Obama should push for transparency in respect of food labelling to help build public trust.

The 'Monsanto Protection Act'

There has been a murky debate over the introduction, passage and subsequent repeal of the Farmer Assurance Provision, 'Monsanto Protection Act'.[157] A large number of civil society groups complained about the measure:

> Earlier this year, hundreds of thousands of Americans called their elected officials to voice their frustration and disappointment over the inclusion of a controversial policy rider (Sec. 735) – dubbed 'the Monsanto Protection Act' – in the Continuing Resolution spending bill (H.R. 933). The rider represents a serious assault on the fundamental safeguards of our judicial system, and could negatively impact farmers, the environment and public health across America. Yet it was quietly slipped into HR 933 without congressional debate, hearings or input from any of the relevant committees.[158]

---

154 Kucinich, E. 2014. Members of Congress, farmers and businesses call on Obama to Fulfill campaign promise on GMO labelling. *The Huffington Post*, 16 January 2014. [Online]. Available at: http://www.huffingtonpost.com/elizabeth-kucinich/post _6676_b_4612219.html [accessed 2 April 2014].

155 Ibid.

156 Ibid.

157 Sheets, C. 2013. 'Monsanto Protection Act' killed in Senate: Controversial measure removed from Spending Bill. *International Business Times*, 27 September 2013. [Online]. Available at: http://www.ibtimes.com/monsanto-protection-act-killed-senate-controversial-provision-removed-spending-bill-1412160 [accessed 2 April 2014].

158 Center for Food Safety. 2013. *129 organizations and companies strongly oppose the Monsanto Protection Act*, 11 September 2013. [Online]. Available at: http://www. centerforfoodsafety.org/files/group-letter-opposing-biotech-rider-in-fy14-cr_14926.pdf [accessed 2 April 2014].

The civil society groups objected that 'Sec. 735 is an unprecedented overstep of the clear-cut boundary of a Constitutionally-guaranteed separation of powers essential to our government.'[159] In their view, 'this rider sets a dangerous precedent for congressional intervention in the judiciary.'[160] The civil society groups maintained that the measure was a significant interference in the operation of the rule of law: 'The ability of courts to review, evaluate and judge an issue that impacts public and environmental health is a strength – not a weakness – of our system.'[161] The groups complained that 'the rider does not merely allow, it would force, the Secretary of Agriculture to immediately grant any requests for permits to allow continued planting and commercialization of an unlawfully approved genetically engineered (GE) crop.'[162] The organisations were concerned that a rubber-stamp process 'could leave public health, the environment and livelihoods at risk'.[163]

Under pressure from the food movement, Democrats Barbara Mikulski and Jeff Merkley of the US Senate took action to remove the 'Monsanto Protection Act' language from the Senate version of the Bill.[164] Monsanto defended the measures.[165] The biotechnology company emphasised that the purpose of the measure was 'to reinforce the integrity of the regulatory system and protect farmers from the disruption of frivolous lawsuits'.[166] The company maintained that there was a need to ensure that the regulatory system was 'based on real science, operating in a timely and data-driven manner to deliver choices to farmers and the economy they support'.[167] Monsanto stressed: 'We will be working to support the efforts of the USDA and US EPA to make the regulatory system work to provide timely, science-based decisions on new products designed to help our farmers produce food and fiber as efficiently and sustainably as possible.'[168]

The Safe and Accurate Food Labeling Act 2014

There has also been discussion as to whether there would be federal efforts to make it illegal for the states to engage in the labelling of GM foods.

In April 2014, US Representative, Mike Pompeo, a Republican congressman from Kansas introduced legislation which would nullify state efforts to require the

---

159   Ibid.

160   Ibid.

161   Ibid.

162   Ibid.

163   Ibid.

164   Sheets, C. 2013, above n 157.

165   Monsanto, 2013. *Expiration of the Farmer Assurance Provision*, 26 September 2013. [Online]. Available at: http://monsantoblog.com/2013/09/26/expiration-of-the-farmer-assurance-provision/ [accessed 2 April 2014].

166   Ibid.

167   Ibid.

168   Ibid.

labelling of GM foods.[169] The legislation was called the *Safe and Accurate Food Labeling Act* 2014.[170] Pompeo explained the federal legislative proposal:

> We've got a number of states that are attempting to put together a patchwork quilt of food labeling requirements with respect to genetic modification of foods. That makes it enormously difficult to operate a food system. Some of the campaigns in some of these states aren't really to inform consumers but rather aimed at scaring them. What this bill attempts to do is set a standard.[171]

Pompeo maintained that GM foods are safe and 'equally healthy' and no special labelling was required or needed. He observed that GM crops: have 'to date made food safer and more abundant' and have 'been an enormous boon to all of humanity'.[172]

The peak body for the biotechnology industry, BIO, applauded the introduction of the legislation.[173] Cathy Enright, BIO's Executive President for Food and Agriculture, contended: 'We also endorse the Safe and Accurate Food Labeling Act as a federal GMO labeling solution that helps to address consumers' questions about GMO safety and provides them with tools necessary to make informed decisions'.[174] The Coalition for Safe Affordable Food was established in order to support the passage of the legislative bill.[175]

In response, Elizabeth Kucinich suggested that Pompeo's bill was designed to 'muddy congressional waters and keep consumers in the dark by preventing GMO labelling.'[176] She suggested that the Grocery Manufacturers Association was the driving force behind Pompeo's proposal. She also wondered whether the Coalition for Safe Affordable Food was really a popular movement, or just an exercise in astroturfing. Kucinich observed: 'The overall strategy is one that

---

169   Gillam, C. 2014. 'U.S. Bill Seeks to Block Mandatory GMO Food Labeling by States,' *Reuters*, 9 April 2014. [Online]. Available at: http://www.reuters.com/article/2014/04/09/usa-gmo-lawmaking-idUSL2N0N115F20140409 [accessed 2 April 2014].

170   *Safe and Accurate Food Labeling Act* of 2014 (HR 4432). Mike Pompeo, 'Pompeo, Butterfield Release Bipartisan Food Labeling Reform,' Press Release, 10 April 2014. [Online]. Available at: http://pompeo.house.gov/news/documentsingle.aspx?DocumentID=376238 [accessed: 2 April 2014].

171   Gillam, C. 2014, above n 169.

172   Ibid.

173   BIO, 2014, 'BIO Applauds Introduction of the Safe and Accurate Food Labeling Act,' *Press Release*, 9 April 2014. [Online]. Available at: http://www.bio.org/media/press-release/bio-applauds-introduction-safe-and-accurate-food-labeling-act [accessed 2 April 2014].

174   Ibid.

175   *Coalition for Safe and Affordable Food.* [Online]. Available at: http://coalitionforsafeaffordablefood.org/ [accessed 2 April 2014].

176   Kucinich, E. 2014. GMO Pushers and The Art of War, *The Huffington Post*, 10 April 2014. [Online]. Available at: http://www.huffingtonpost.com/elizabeth-kucinich/gmo-labeling_b_5120692.html [accessed 2 April 2014].

Sun Tzu could be proud of: obfuscation'.[177] She retorted: 'Tired and false claims about GMOs saving the world are not a sufficient reason to deny citizens their right to know'.[178]

There may be insufficient support for this bill under the present US Congress. If the bill is passed, there could well be conflict between state and federal regimes in respect of GM food labelling.

Trade, Investment, and GM Food Labelling

There has been significant debate over the Obama administration's pursuit of regional trade agreements – such as the *Trans-Pacific Partnership* (TPP), involving a dozen countries in the Pacific Rim,[179] and the *Trans-Atlantic Trade and Investment Partnership*, a proposed trade agreement between the US and the European Union. Will the *Trans-Pacific Partnership* and the *Trans-Atlantic Trade and Investment Partnership* affect policy flexibilities in respect of the public governance of food, the environment and public health? There has been concern about the impact of such regional trade initiatives on packaging and labelling laws and regulations – such as graphic health warnings and the plain packaging of tobacco products, food nutrition information and GM food labelling.

There has been significant opposition to the fast-tracking of the *Trans-Pacific Partnership* and the *Trans-Atlantic Trade and Investment Partnership*. A grand coalition of civil society organisations – including campaigners on GM food labelling – has lobbied the US Congress against fast-tracking the trade deals.[180] The Organic Consumers Association has opposed the grant of a fast track authority because it is concerned that secret trade agreements threaten food safety and subvert democracy: 'If these deals are rammed through Congress without scrutiny or debate, we could lose our right to regulate factory farms and GMOs.'[181] LabelGMOs.org has also opposed Fast Track: 'We believe in food sovereignty for all people and are taking a strong stand against corporate control of our food supply.'[182] GMO Inside

---

177   Ibid.

178   Ibid.

179   Brewster, K. 2013. Trans-Pacific Partnership could damage Australia, *Lateline*, ABC, 10 October 2013. [Online]. Available at: http://www.abc.net.au/lateline/content/2013/s3866749.htm [accessed 2 April 2014].

180   *Stop Fast Track*. [Online]. Available at: http://www.stopfasttrack.com/ [accessed 2 April 2014].

181   Ibid. See Organic Consumers Association. *Don't let Congress 'fast-track' dangerous trade deals*. [Online]. Available at: http://salsa3.salsalabs.com/o/50865/p/dia/action3/common/public/?action_KEY=12779 [accessed 2 April 2014].

182   *Stop Fast Track*. [Online], above n 180. See also Hall, S. 2013. *Monsanto to outlaw GM labelling worldwide through secret trade – the TPP*, 26 November 2013. [Online]. Available at: http://www.labelgmos.org/monsanto_to_outlaw_gmo_labeling_worldwide_through [accessed 2 April 2014].

opposes Fast Track because of its concern that 'under the TPP GMO labels for US food would not be allowed.'[183]

GMO Free Arizona fears that the 'TPP will pre-empt important GMO labelling and moratoriums'[184] and is concerned that 'the TPP would unravel our movement's work with GMO labelling, GMO cultivation bans and gut food and environmental safety standards.'[185] In addition to such specific concerns about GM food labelling, there are broader concerns about how the trade deals will affect intellectual property, public health, the environment, consumer rights and workers' jobs and wages.

The debate over GM food labelling is a highly polarised discussion – even in international discussions over the *Trans-Pacific Partnership* and the *Trans-Atlantic Trade and Investment Partnership*. It is still hard to determine whether such hopes or fears are justified, in the absence of open, public text and negotiating positions.

There has been much debate about the secrecy of such regional trade agreements. Critics have lamented the lack of transparency, accountability, legislation and public participation. US Congressional Democrat Senator Elizabeth Warren warned of the dangers of the *Trans-Pacific Partnership* and the *Trans-Atlantic Trade and Investment Partnership*:

> For big corporations, trade agreement time is like Christmas morning. They can get special gifts they could never pass through Congress out in public. Because it's a trade deal, the negotiations are secret and the big corporations can do their work behind closed doors. We've seen what happens here at home when our trading partners around the world are allowed to ignore workers' rights, wages, and environmental rules. From what I hear, Wall Street, pharmaceuticals, telecom, big polluters, and outsourcers are all salivating at the chance to rig the upcoming trade deals in their favor.[186]

She commented: 'I believe that if people would be opposed to a particular trade agreement, then that trade agreement should not happen.'[187] The Democrats in the US Congress have been reluctant to provide President Barack Obama with a fast-track authority in respect of the regional trade deals.[188] As such, there is

---

183 *Stop Fast Track.* [Online], above n 180. See also GMO Inside.org. [Online]. Available at: http://gmoinside.org/ [accessed 2 April 2014].

184 *Stop Fast Track.* [Online], above n 180.

185 Ibid.

186 Warren, E. 2013. *Remarks to the AFL-CIO Convention*, 8 September 2013. [Online]. Available at: http://www.warren.senate.gov/?p=press_release&id=234 [accessed 2 April 2014].

187 Ibid.

188 Needham, V. 2014. Pelosi comes out against fast track. *The Hill*, 12 February 2014. [Online]. Available at: http://thehill.com/homenews/house/198297-pelosi-comes-out-against-fast-track-bill [accessed 2 April 2014].

an impasse between the Obama administration and the US Congress over these sweeping trade deals, spanning the Pacific Rim and the Atlantic.

The Trans-Pacific Partnership

There has been much controversy over the *Trans-Pacific Partnership* – a plurilateral trade agreement involving a dozen nations from throughout the Pacific Rim.[189] One of the most contentious areas of debate has been the question of agriculture. Deborah Elms comments that there has been discussion over the inclusion of agriculture in the deal:

> In preparing their calculations about the net benefits of the TPP, many officials realised that if agricultural trade were excluded from the final agreement (or if significant sectors were carved out of the final document, the net economic benefits from the TPP would be lower. Because some agricultural sectors had not been liberalised or had not been fully liberalised in past agreements, there was still scope for improvement in the TPP ... If any one area could be carved out as too sensitive for inclusion, it would establish the possibility that countries could carve out other highly sensitive issues from the text elsewhere.[190]

There are a range of agricultural issues under debate – including tariffs and harmonised system codes, rules of origin, sanitary and phytosanitary rules, intellectual property standards, investment, the protection of the environment and the use of regulations, such as food labelling.

In November 2013 WikiLeaks published a Draft Text of the Intellectual Property chapter of the *Trans-Pacific Partnership*.[191] The Intellectual Property chapter includes text on patent law, trade mark law, copyright law, data protection and intellectual property enforcement.[192] A number of the US proposals are

---

189   See Voon, T. (ed.). 2013. *Trade liberalisation and international co-operation: A legal analysis of the Trans-Pacific Partnership Agreement*, Cheltenham (UK) and Northampton (Mass.): Edward Elgar; Lim, C., Elms, D. and Low, P. (eds). 2012. *The Trans-Pacific Partnership: A quest for a Twenty-First Century trade agreement*, Cambridge: Cambridge University Press; Kelsey, J. (ed.). 2010. *No Ordinary Deal: Unmasking the Trans-Pacific Partnership Free Trade Agreement*. Wellington: Bridget Williams Books Inc; Kelsey, J, 2013, *Hidden Agendas: What we need to Know About the TPPA*, Wellington: Bridget Williams Books Inc.

190   Elms, D. 2013. Agriculture and the Trans-Pacific Partnership, in *Trade Liberalisation and International Co-operation: A Legal Analysis of the Trans-Pacific Partnership Agreement* edited by T. Voon. Cheltenham (UK) and Northampton (Mass.): Edward Elgar, 106–30, 107.

191   WikiLeaks. 2013. *Secret Trans-Pacific Partnership Agreement: The IP chapter*, 13 November 2013. [Online]. Available at: https://wikileaks.org/tpp/ [accessed 22 May 2014].

192   Rimmer, M. 2013. Our future is at risk: Disclose the Trans-Pacific Partnership now. *New Matilda*, 15 November 2013. [Online]. Available at: https://newmatilda.com/2013/11/15/our-future-risk-disclose-tpp-now [accessed 2 April 2014].

particularly about boosting the intellectual property rights of agricultural companies, the biotechnology industry and the food industry. In January 2014 WikiLeaks also published a Draft Text of the Environment Chapter of the *Trans-Pacific Partnership*.[193] The text reveals a weak regime for the protection of the environment, biodiversity and climate change.[194] As such, the Environment Chapter will do little to provide for protection of public regulation in respect of food and the environment. There has also been much concern about the proposals in respect of the Investment Chapter.[195] The investor-state dispute settlement regime would enable foreign investors to bring tribunal action against nation states in respect of government decisions that adversely affect their foreign investments.[196]

The Biotechnology Industry Organization (BIO) has maintained that the status quo in the US should be a model for the *Trans-Pacific Partnership*:

> With regard to labelling of foods derived from agricultural biotechnology, BIO recommends the development of labelling practices consistent with the U.S. Food and Drug Administration (FDA) Draft Guidance. Therefore, any mandatory or required labelling for genetically engineered products should be science based, such as if the product has been significantly changed nutritionally or if there have been changes in other significant health-related characteristics of the food (allergenicity, toxicity, or composition). Voluntary labelling should be truthful and not misleading.[197]

BIO has maintained:

> The US government has stated the intention to treat the *Trans-Pacific Partnership* as a model agreement for the 21st century, and therefore BIO believes that sound, objective and science-based approaches to agricultural biotechnology regulation

---

193   WikiLeaks. 2013. *WikiLeaks release of secret Trans-Pacific Partnership: Environment chapter consolidated text*, 24 November 2013. [Online]. Available at: https://wikileaks.org/tpp-enviro/ [accessed 2 April 2014].

194   Rimmer, M. and Wood, C. 2014. Trans-Pacific Partnership greenwashes dirty politics. *New Matilda*, 17 January 2014. [Online]. Available at: https://newmatilda.com/2014/01/16/tpp-greenwashes-dirty-politics [accessed 2 April 2014].

195   Rimmer, M. 2012. A dangerous investment: Australia, New Zealand, and the Trans-Pacific Partnership. *The Conversation*, 2 July 2012. [Online]. Available at: http://theconversation.edu.au/a-dangerous-investment-australia-new-zealand-and-the-trans-pacific-partnership-7440 [accessed 2 April 2014].

196   United Nations Conference on Trade and Development (UNCTAD), 2014, *'Recent Developments in Investor-State Dispute Settlement: Updated for the Multilateral Dialogue on Investment'*, April 2014. [Online]. Available at: http://unctad.org/en/PublicationsLibrary/webdiaepcb2014d3_en.pdf [accessed 2 April 2014].

197   Biotechnology Industry Organization, 'BIO comments on the proposed accession of Malaysia to the *Trans Pacific Partnership* (TPP) negotiations,' 22 November 2010. [Online]. Available at: http://www.bio.org/node/228 [accessed 2 April 2014].

should be a top priority, particularly with respect to the challenges facing global agriculture and energy supplies in the 21st century and beyond.[198]

The biotechnology industry is thus keen for the *Trans-Pacific Partnership* to address regulatory restrictions in respect of agricultural biotechnology – including in respect of labelling. Accordingly, as James Trimarco has warned: 'The *Trans Pacific Partnership* is likely to be a setback for efforts to regulate and label GMO foods.'[199]

There has been concern about the impact of the *Trans-Pacific Partnership* on food regulation. Sharon Friel, Deborah Gleeson, and Libby Hattersley have stressed that 'international trade agreements bring new transnational food companies into countries, along with new food advertising and promotion.'[200] The health and trade researchers observed:

> The *Trans Pacific Partnership* is likely to provide stronger investor protections and enable greater (food) industry involvement in policy-making. It could lead to sweeping changes to domestic regulatory systems, and open up new opportunities for companies to appeal against domestic policies they consider to be a violation of their privileges under the agreement. Together, these changes would weaken the ability for governments to protect public health by, for example, limiting imports and domestic manufacturing of unhealthy foods and drinks.[201]

There has been particular disquiet that 'at the 15th round of negotiations in Auckland last December, the Malaysian government – supported by the US – reportedly suggested restricting the amount of information food companies would be required to provide about ingredients and formulae of processed food products.'[202] Friel and her collaborators comment that 'these sorts of proposals raise concerns about consumer access to information about food products, as well as the ability of governments to regulate food labelling on public health grounds.'[203] The group maintained: 'Measures like that one will undermine health policy goals and extend the control of the food industry over domestic policy.'[204] In their view, 'Re-balancing food industry influence in the negotiation process with input from the health sector

---

198   Ibid.

199   Trimarco, J. 2013. Will a secretive international trade deal ban GMO labelling? *Yes Magazine!*, 18 October 2013. [Online]. Available at: http://www.yesmagazine.org/planet/will-secretive-international-trade-deal-ban-gmo-labeling-trans-pacific-partnership [accessed 2 April 2014].

200   Friel, S., Gleeson, D. and Hattersley, L. 2013. Trans Pacific Partnership puts member countries' health at risk. *The Conversation*, 9 May 2013. [Online]. Available at: http://theconversation.com/trans-pacific-partnership-puts-member-countries-health-at-risk-13711 [accessed 2 April 2014].

201   Ibid.

202   Ibid.

203   Ibid.

204   Ibid.

is vital.'[205] Friel and her collaborators called for a greater focus upon the protection of public health and nutrition in the trade negotiations: 'Public health advocates and health policymakers must engage with trade negotiations to preserve policy space for public health goals before the window of opportunity closes.'[206]

Barbara Chicherio complained that the *Trans-Pacific Partnership* was a boon to Monsanto and would undermine public regulation in respect of food and health.[207] She was particularly worried about the labelling of GM foods:

> The labelling of foods containing GMOs (Genetically Modified Organisms) will not be allowed. Japan now has labelling laws for GMOs in food. Under the TPP Japan would no longer be able to label GMOs. This situation is the same for New Zealand and Australia. The US is just beginning to see some progress towards labelling GMOs. Under the TPP GMO labels for US food would not be allowed.[208]

The Sustainability Council in New Zealand has also been particularly concerned about the impact of the Trans-Pacific Partnership upon GM Food Labelling. In an opinion-editorial for *The New Zealand Herald*, Stephanie Howard and Simon Terry wondered: 'Will losing the right to choose GM-free food be a price of the next and biggest free trade deal?'[209] The researchers at the Sustainability Council observed:

> The United States has made clear that a priority for the proposed Trans Pacific Partnership (TPP) is the abolition of laws that require genetically modified foods to be labelled. That puts New Zealand in its sights because GM ingredients in food products must generally be labelled here.

> Although there are exemptions such as highly refined oils and GM contamination below 1 per cent, New Zealand food companies and supermarkets have avoided ingredients in their products that would trigger the labelling and retailers essentially do not stock products tagged as GM.

> Without the labelling law, New Zealanders who want to avoid genetically modified food would have to rely on the willingness of producers to declare such content – or a patchwork of independent testing.

---

205 Ibid.

206 Ibid.

207 Chicherio, B. 2013. How new 'free' trade deal aids Monsanto. *Green Left Weekly*, 28 September 2013. [Online]. Available at: https://www.greenleft.org.au/node/55054 [accessed 2 April 2014].

208 Ibid.

209 Howard, S. and Terry, S. 2011. 'Let's Insist on Labels for GM Food', *The New Zealand Herald*, 10 November 2011. [Online]. Available at: http://www.nzherald.co.nz/opinion/news/article.cfm?c_id=466&objectid=10764893 [accessed 2 April 2014].

Loss of the right to know when a product contains GM ingredients could quickly slide into effective loss of the right to choose everyday foods that are not genetically modified. Instead of it being the norm for food companies to strive to keep GM out of their products, this could become the preserve of niche eco brands.[210]

The writers alleged: 'The reason Washington wants to stamp out all mandatory labelling is plain: the US is the world's largest producer of GM crops and its soy and corn are now almost all genetically modified.'[211]

Olivier De Schutter, the United Nations special rapporteur on the right to food, and Kaitlin Cordes, a food security researcher from Columbia University, have made an important contribution to the policy debate over the *Trans-Pacific Partnership*.[212] The writers lament the failure to consider the human rights implications of the agreement:

> Whether trade liberalization generally helps or harms the most vulnerable is a complex question. But that theoretical debate should not prevent us from carrying out a thorough human-rights impact assessment on the terms of the deal currently on the table. Such an assessment should be conducted before the TPP negotiations reach any final agreement on the relevant issues, and it should not overlook how the terms are implemented in practice. Unfortunately, TPP member states have not only failed to do this; they have also excluded independent organizations from the assessment process by refusing to provide access to draft texts.[213]

Citing the work of Joseph Stiglitz, Olivier De Schutter and Kaitlin Cordes worry that 'the TPP's emphasis on regulatory policies suggests that business interests will trump human rights.'[214]

In particular, De Schutter and Cordes express concerns about the impact of the *Trans-Pacific Partnership* upon farming, agriculture and food security:

> Leaked drafts of intellectual-property proposals show an obstinate US effort to require patent protections for plants and animals, thus going beyond the World Trade Organization's *TRIPS Agreement* 1994. The US stance could further

---

210   Ibid.

211   Ibid.

212   O. De Schutter and K. Cordes, 'Trading Away Human Rights', *Project Syndicate*, 7 January 2014. [Online]. Available at: http://www.project-syndicate.org/commentary/olivier-de-schutter-and-kaitlin-y--cordes-demand-that-the-trans-pacific-partnership-s-terms-be-subject-to-a-human-rights-impact-assessment#9Lq5cFjsOIfGZhhf.99 [accessed 2 April 2014].

213   Ibid.

214   Ibid.

restrict farmers' access to productive resources, thus affecting the right to food. And such proposals would limit governments' options when addressing wider food-related human-rights issues.[215]

The writers warn: 'This clash of interests contravenes basic principles of international law, namely that countries' trade deals must not conflict with their obligations under human-rights treaties.'[216] The policy-makers emphasised: 'That is why a human-rights impact assessment must be conducted – and necessary additional safeguards added – before any TPP deal is signed.'[217]

De Schutter and Cordes stressed that transparency and inclusiveness should be the prerequisites of any deal: 'Although trade negotiations require discretion to avoid political grandstanding by participants, the secrecy that currently surrounds the TPP talks is preventing important human-rights arguments from being aired.'[218] De Schutter and Cordes emphasised that a change in the process could address significant injustices: 'If they truly want the TPP to be a model for the twenty-first-century global economy, as they claim, then they should show real leadership.'[219] The pair advised: 'The TPP negotiators should consider the rights of everyone affected by the deal and act in the public interest, not just the special interests of the economic players that stand to benefit the most.'[220]

In a report to the United Nations on the right to food, Olivier De Schutter has explored the interaction between intellectual property and food security.[221] The Special Rapporteur argued that:

> in order to ensure that the development of the intellectual property rights regime and the implementation of seed policies at the national level are compatible with the right to food, States should ... support efforts by developing countries to establish a sui generis regime for the protection of intellectual property rights which suits their development needs and is based on human rights.[222]

Olivier De Schutter was hopeful that democracy and diversity could help mend broken food systems.[223] He observed: 'The greatest deficit in the food economy

---

215 Ibid.
216 Ibid.
217 Ibid.
218 Ibid.
219 Ibid.
220 Ibid.
221 De Schutter, O. 2014. *The Transformative Potential of the Right to Food: Report of the Special Rapporteur on the Right to Food*, Human Rights Council, United Nations General Assembly, A/HRC/25/57, 24 January 2014. [Online]. Available at: http://www.srfood.org/images/stories/pdf/officialreports/20140310_finalreport_en.pdf [accessed 2 April 2014].
222 Ibid., 21–2.
223 De Schutter, O. 2014. Democracy and diversity can mend broken food systems – Final diagnosis from UN Right to Food expert. *United Nations*, 10 March 2014. [Online].

is the democratic one.'[224] And argued: 'By harnessing people's knowledge and building their needs and preferences into the design of ambitious food policies at every level, we would arrive at food systems that are built to endure.'[225] He maintained that 'food democracy must start from the bottom-up, at the level of villages, regions, cities, and municipalities.'[226] Olivier De Schutter has insisted that there is a need for an 'alternative paradigm for the 21st century': 'There is much that can be done by developing countries themselves to support small-scale farmers with the land, credit, technology and market access they need.'[227]

The Trans-Atlantic Trade and Investment Partnership

In addition to the Trans-Pacific Partnership, President Barack Obama has also been pursuing a Trans-Atlantic Trade and Investment Partnership.[228]

In 2013, Utah Republican Senator Orrin Hatch and Montana Democrat Senator Baucus wrote to the US Trade Representative on the priorities that should be embodied in any US-European Trade Agreement.[229] The pair demanded access for US agricultural exports, a relaxation of biotechnology regulations and strong intellectual property protection. Hatch and Baucus insisted that 'broad bipartisan Congressional support for expanding trade with the EU depends, in large part, on lowering trade barriers for American agricultural products.'[230] The pair lamented: 'The EU has historically imposed sanitary and phytosanitary measures that act as significant barriers to US-EU trade, including the EU's restrictions on genetically engineered crops.'[231] Hatch and Baucus also stressed that 'Congressional support will also require strong intellectual property protection.'[232] The pair noted: 'Intellectual property is America's competitive advantage, underpinning a wide range of industries including manufacturing, food processing, information and

---

Available at: http://www.srfood.org/en/democracy-and-diversity-can-mend-broken-food-systems-final-diagnosis-from-un-right-to-food-expert [accessed 2 April 2014].

224   Ibid.

225   Ibid.

226   Ibid.

227   De Schutter, O. 2014. Ending hunger – the rich world holds the keys. *The Ecologist*, 25 March 2014. [Online]. Available at: http://www.theecologist.org/blogs_and_comments/commentators/2333245/ending_hunger_the_rich_world_holds_the_keys.html [accessed 2 April 2014].

228   *Trans-Atlantic Trade and Investment Partnership*. [Online]. Available at: http://ec.europa.eu/trade/policy/in-focus/ttip/ [accessed 2 April 2014].

229   Baucus, M. and Hatch, O. 2013. Baucus, Hatch outline priorities for potential US-EU trade agreement', Press Release, *The United States Committee on Finance*, 12 February 2013. [Online]. Available at: http://www.finance.senate.gov/newsroom/chairman/release/?id=17b2fd73–067d-4a4a-a50f-a00265efbf67 [accessed 2 April 2014].

230   Ibid.

231   Ibid.

232   Ibid.

communications technology, entertainment, biotech, pharmaceuticals and financial services.'[233] In their view: 'It is imperative that US trade agreements protect US innovation and allow our innovative industries to compete in global markets.'[234]

The Food Democracy Now! movement has been alarmed by the influence of biotechnology companies on the development of the *Trans-Pacific Partnership* and the *Trans-Atlantic Trade and Investment Partnership*: 'Just as we are on the verge of winning GMO labelling in the US, these secret negotiations could eliminate the ability of countries around the world to label GMOs or impose common sense restrictions on the sale of genetically engineered seed and food in their countries if Monsanto and other biotech companies get their way.'[235]

The European Commission has sought to deny that the European Union will be forced to change its laws on GMOs.[236] The Commission insisted: 'Basic laws, like those relating to GMOs or which are there to protect human life and health, animal health and welfare, or environment and consumer interests will not be part of the negotiations.'[237] The Commission observed:

> Under EU rules, GMOs that have been approved for use as food, for animal feed or for sowing as crops can already be sold in the EU. Applications for approval are assessed by the European Food Safety Authority (EFSA) and then sent to EU Member States for their opinion. So far, 52 GMOs have been authorised. The safety assessment which EFSA carries out before any GMO is placed on the market and the risk management procedure will not be affected by the negotiations.[238]

The Commission agreed that there would be co-operation on the regulation on biotechnology, emphasising that 'The EU and US already exchange information on policy, regulations and technical issues concerning GMOs.' and that 'cooperation of this sort helps minimise the effect on trade of our respective systems for approving GMOs.'[239]

However, the Directorate General for Internal Policies at the European Parliament has highlighted wide divisions between the European Union and the US on the regulation of the environment.[240] The study highlighted the significant prescriptions by the European Union in respect of GM food labelling:

---

233 Ibid.

234 Ibid.

235 Food Democracy Now! *Stop the Secret Trade Deals: The Monsanto Protection Act on Steroids*. [Online]. Available at: http://action.fooddemocracynow.org/sign/stop_tpp_ tafta_monsanto_protection_act_on_steroids/ [accessed 2 April 2014].

236 European Commission. *The Trans-Atlantic Trade and Investment Partnership*. [Online]. Available at: http://ec.europa.eu/trade/policy/in-focus/ttip/ [accessed 2 April 2014].

237 Ibid.

238 Ibid.

239 Ibid.

240 Directorate General for Internal Policies at the European Parliament. 2013. *Legal Implications of TTIP for the Acquis Communautaire in ENVI Relevant Sectors*, European

Regulation No 1829/2003 also requires the labelling of GMOs and products thereof for food use when they contain, consist of, or are produced from GMOs in a proportion higher than 0.9 per cent of the food ingredients considered individually or food consisting of a single ingredient (Art 12(2)). The same applies for feed containing material where the proportion of GMOs is higher than 0.9 per cent of the feed and of each feed of which it is composed (Art 24(2)). If the proportion is less than 0.9%, labelling is not required, provided that the presence of the GMO is adventitious or technically unavoidable. Regulation (EC) No 1830/2003 sets specific requirements for the traceability of GMOs at each stage of production and placing on the market, with monitoring of labelling and of the potential effects on human health or the environment.[241]

By contrast, the report observed that 'the US policy framework can be understood as the product of an ambivalent institutional mission: in the case of food and food crops, for instance, the FDA has sought simultaneously to ensure food safety and to promote biotechnology in agriculture.'[242]

The report noted: 'While in the EU there is a specific legal framework to regulate GMOs, whether for food and feed uses or for cultivation, the applicable US legal and regulatory framework is comparatively basic and limited to non-binding policy statements on the application of existing product-related laws to GMOs.'[243] The report stressed: 'In addition, while labelling of GMOs and products thereof for food use is mandatory under EU legislation (if the proportion of GMOs is higher than 0.9 per cent), it is currently only voluntary under US federal law, although there are early initiatives at the state level to introduce mandatory labelling for improved consumer information.'[244] The study recommended: 'In light of the highly differing EU and US regulations applicable to GMOs, any TTIP provisions which could apply to GMOs should be carefully reviewed, in order not to inadvertently undermine the stricter EU standards for the authorisation of GMOs (e.g., risk assessment), as well as transparency of GMO-related information (notably public consultation, registration and labelling).'[245]

Michael Lipsky, a senior fellow at Demos, has wondered, 'Will European requirements for labelling GMO foods survive new trade negotiations?'[246] He feared that US and European negotiators could 'bargain away a key element in

---

Parliament. [Online]. Available at: http://www.europarl.europa.eu/RegData/etudes/etudes/join/2013/507492/IPOL-ENVI_ET%282013%29507492_EN.pdf [accessed 2 April 2014].

241   Ibid., 34–5.
242   Ibid., 38.
243   Ibid., 23.
244   Ibid.
245   Ibid., 27.
246   Lipsky, M. 2013. Will European requirements for labelling for GMO foods survive new trade negotiations? *The Huffington Post*, 3 July 2013. [Online]. Available at: http://www.huffingtonpost.com/michael-lipsky/will-european-requirement_b_3535795.html [accessed 2 April 2014].

American resistance to GMO foods'.[247] Lipsky worried that 'the proposed *Trans-Atlantic Trade and Investment Partnership* (TTIP), also referred to as a *Trans-Atlantic Free Trade Agreement* (TAFTA), will focus on "normalizing" regulatory practices that business interests deem limit trade, including the European approach to genetically modified foods.'[248] Lipsky commented:

> In international trade negotiations, most often Americans worry that domestic regulatory protections or special tariffs will be bargained away in the interests of increasing trade overall. In the case of GMO foods, many Americans have exactly the opposite interest – maintaining the regulatory protections of foreign trade partners so that the precautionary approach to GMO foods practiced by Europeans remains intact.[249]

Lipsky stressed that the European experience was an important counterpoint: 'For Americans who believe that people should be able to know whether their food has been genetically modified, the European experience is a critical reference point.'[250] He noted: 'The cautionary European policies toward GMO foods represent clear and tested alternative approaches to bio-engineering the food supply.'[251] Lipsky observed: 'It is difficult for opponents of GMO food labelling to marginalize their opponents when virtually every advanced industrial country except the US (64 according to a recent count) requires labelling and subscribes to restrictive GMO policies.'[252]

There has been much debate over the proposal for the inclusion of an investor-state dispute settlement clause in the *Trans-Atlantic Trade and Investment Partnership*. Ska Keller, an MEP with the European/EFA Group has been concerned that investment clauses could be deployed against a range of public regulations:

> States are often held back from adopting regulations solely by the threat of lawsuits – this is called the 'chilling effect'. If an action is pending, host countries often act in line with previous cases, which includes withdrawing regulations out of fear of having to pay out high levels of compensation later on.[253]

Keller is particularly concerned about US companies deploying investment chapters against European environmental laws. She questions: 'Should we upend all the democratic decision-making processes in Europe and cause such a legal

---

247   Ibid.
248   Ibid.
249   Ibid.
250   Ibid.
251   Ibid.
252   Ibid.
253   Keller, S. 2014. 'Investor-state lawsuits threaten democracy', *TTIP: Beware what lies beneath*, 26 March 2014. [Online]. Available at: http://ttip2014.eu/blog-detail/blog/Ska%20Keller%20Investors%20TTIP.html [accessed 2 April 2014].

shake-up just for provisions like [investor-state dispute settlement clauses] and the few European companies who had problems with US American courts?'[254] Her preference is to exclude investor-state dispute settlement clauses from the *Trans-Atlantic Trade and Investment Partnership* altogether.

Jose Bove – an MEP with the European Green/EFA group – has emphasised the need to defend the right to healthy, safe food in the negotiations over the *Trans-Atlantic Trade an Investment Partnership*.[255] He warned that the proposed deal could impact public regulations:

> We believe this is an assault on our democratic right to regulate. It could majorly impair our democratic institutions ability to legislate by creating a 'chilling effect' on proposals aimed at enhancing the public good, but that go against the interests of private companies trading in both regions. It could impact the extent to which our food is labelled or what processes become acceptable in the making of our food, among many. The Greens will not accept any trade deal that undermines democratic institutions' right to regulate.[256]

Bove maintained: 'It is time for all of us, our farming community, our citizens, and anyone who enjoys safe and healthy food, to become aware of what is being negotiated away- and to do everything in our power to stop it.'[257]

There has been much controversy over the possibility of the inclusion of an investor-state dispute settlement regime in the *Trans-Atlantic Trade and Investment Partnership*.[258]

Glyn Moody has observed that there has been a backlash against the inclusion of investor-state dispute settlement, around the world.[259] He noted that Germany

---

254   Ibid.

255   Bove, J. 2014. We defend the right to healthy, safe food: We won't trade this away in TTIP, *TTIP: Beware what lies beneath*, 26 March 2014. [Online]. Available at: http://ttip2014.eu/blog-detail/blog/Food%20and%20agriculture.html [accessed 2 April 2014].

256   Ibid.

257   Ibid.

258   European Commission, 2014 *'Commission to consult European public on provisions in EU-US trade deal on investment and investor-state dispute settlement'*, 21 January 2014. [Online]. Available at: http://europa.eu/rapid/press-release_IP-14–56_en.htm [accessed 2 April 2014]; St. Louis, M. 2014, 'Public Interest Critique of ISDS: Drastic Increase in Government Liability', *Public Citizen's Global Trade Watch*, 17 March 2014; Kleinheisterkamp, J. *'Is there a Need for Investor-State Arbitration in the Transatlantic Trade and Investment Partnership (TTIP)?'* (14 February 2014). [Online]. Available at: http://ssrn.com/abstract=2410188 [accessed 2 April 2014]; Moody, G., 2014, Even the German Government Wants Corporate Sovereignty out of TAFTA/ TTIP, *TechDirt*, 17 March 2014. [Online]. Available at: http://www.techdirt.com/articles/20140313/10571526568/even-german-government-wants-corporate-sovereignty-out-taftattip.shtml [accessed 2 April 2014].

259   Moody, G. 2014. Corporate sovereignty provisions called into question around the world. *TechDirt*, 26 March 2014. [Online]. Available at: https://www.techdirt.com/

has called for the repudiation of investment clauses in the *Trans-Atlantic Trade and Investment Partnership*.

The British environmentalist George Monbiot has written a series of articles about the *Trans-Atlantic Trade and Investment Partnership*.[260] He raised concerns about the nature of the trade deal, suggesting that it had been captured by transnational corporate elites. Monbiot was particularly concerned about the possibility of the inclusion of an investment chapter in the *Trans-Atlantic Trade and Investment Partnership*.[261] Monbiot observed:

> Investor-state rules could be used to smash any attempt to save the NHS from corporate control, to re-regulate the banks, to curb the greed of the energy companies, to renationalise the railways, to leave fossil fuels in the ground. These rules shut down democratic alternatives. They outlaw leftwing politics.[262]

Monbiot was concerned that the treaty would grant big business the remarkable ability 'to sue the living daylights out of governments which try to defend their citizens'.[263] He warned that the regime 'would allow a secretive panel of corporate lawyers to overrule the will of parliament and destroy our legal protections'.[264]

Conclusion

There is a strong case for GM food labelling – particularly in terms of promoting consumer rights in respect of the right to know about food. Ideally, there should be common standards in respect of GM food labelling. However, GM food labelling

---

articles/20140325/10074526680/corporate-sovereignty-provisions-called-into-question-around-world.shtml [accessed 2 April 2014].

260 Monbiot, G. 2013. From Obamacare to trade, supervision not subversion is the new and very real threat to the state. *The Guardian*, 15 October 2013. [Online]. Available at: http://www.theguardian.com/commentisfree/2013/oct/14/obamacare-trade-superversion-subversion-threat-state [accessed: 2 April 2014]; Monbiot, G. 2013. The Trans-Atlantic Trade deal is a full-frontal assault on democracy. *The Guardian*, 5 November 2013. [Online]. Available at: http://www.theguardian.com/commentisfree/2013/nov/04/us-trade-deal-full-frontal-assault-on-democracy [accessed 2 April 2014]; Monbiot, G. 2013. The lies behind this transatlantic trade deal. *The Guardian*, 3 December 2013. [Online]. Available at: http://www.theguardian.com/commentisfree/2013/dec/02/transatlantic-free-trade-deal-regulation-by-lawyers-eu-us [accessed: 2 April 2014]; and Monbiot, G. 2014. Give and take in the EU-US trade deal? Sure. We give, the corporations take. *The Guardian*, 11 March 2014. [Online]. Available at: http://www.theguardian.com/commentisfree/2014/mar/10/eu-us-trade-deal-give-corporations-take [accessed 2 April 2014].

261 Monbiot, G. 2013, Trans-Atlantic Trade deal, above n 260.

262 Ibid.

263 Ibid.

264 Ibid.

has been a flashpoint for policy conflict in the US between consumer groups, environmentalists, the farm movement, the food industry and the biotechnology industry. The battles have ranged across local politics, federal congressional debate and international trade disputes. As a result, the state of the law is rather fragmented and fractured. There has been a strong grassroots movement in the US, pushing for state initiatives in respect of GM food labelling, particularly through legislative bills and citizen initiated ballots. The food industry and biotechnology industry have sought to thwart such efforts, engaging in lobbying, litigation and well-funded public advertising campaigns. There has also been a significant push at a federal level to legislate in respect of GM food labelling. However, there has been a significant amount of inertia from the Obama administration and a lack of consensus in the US Congress. There has also been considerable debate as to how the state and federal GM food labelling initiatives in the US will be affected by regional trade agreements. Professor Joseph Stiglitz has warned that 'there are trade proposals in the works that threaten to put most Americans on the wrong side of globalization.'[265] There has been a concerted push by the Obama Administration to build regional partnerships with the *Trans-Pacific Partnership* and the *Trans-Atlantic Trade and Investment Partnership*. There has been much debate as to whether such trade agreements will directly or indirectly undermine public regulation in respect of food labelling – particularly in respect of GM food labelling. There is a need to ensure that consumer rights – particularly in respect of food labelling – are properly respected and recognised in such regional and international trade negotiations.

---

265   Stiglitz, J. 2014. On the wrong side of globalization. *The New York Times*, 15 March 2014. [Online]. Available at: http://opinionator.blogs.nytimes.com/2014/03/15/on-the-wrong-side-of-globalization/ [accessed 2 April 2014].

Chapter 8

Unnaturally Natural: Inventing and Eating Genetically Engineered AquAdvantage® Salmon and the Paradox of Nature

Jay Sanderson and Fran Humphries

Introduction

What image of genetically engineered animals do you subscribe to? Are they a frankenfood that have an unnatural origin and violate boundaries between species?[1] Or is genetic engineering just another tool or process for expressing desirable traits at the cellular level, leaving genetically engineered animals no different to animals occurring naturally.[2] While numerous distinctions arise in the debates over genetically engineered animals – including toxic/non-toxic, safe/unsafe and ethical/unethical – it is the question of whether genetically engineered animals are natural that is, perhaps, vexing and most frequently contested. One of the areas in which the question of whether genetically engineered animals are natural is important is in the legal and regulatory frameworks associated with genetically engineered animals. Activists, governments and multinational food corporations, for example, continue to clash over the classification, labelling and marketing of products that contain genetically engineered organisms as 'natural'[3]

---

1   See generally, for example, Tokar, B. 2001. *Redesigning Life? The worldwide challenge to genetic engineering*. London: Zed Books.

2   See for example, Miller, H. and Conko, G. 2004. *The Frankenfood Myth: How protest and politics threaten the biotech revolution*. Westport: Praeger Publishers Inc, 5 (where the authors suggest that 'wide crosses, whose fruits we enjoy every day, are at least as "unnatural" as gene-splicing – if by that term one means a process that does not occur in nature').

3   The issue of labelling genetically engineered animals as 'natural' has resulted in proposed United States federal and state legislation that has been dubbed the 'Right to Know' Acts, which have the goal of requiring the mandatory labelling of genetically engineered foods in the United States. Issues regarding labelling are beyond the scope of this chapter but see Matthew Rimmer, '*Just Label It*: Consumer Rights, GM Food Labelling and International Trade' in this collection. The content and tone of the debates can also be found in Strom, S. 2013. Group seeks special label for food: 'Natural', *The New York Times*, 19 December 2013. [Online]. Available at: http://www.nytimes.com/2013/12/20/business/trade-group-seeks-natural-label-on-modified-food.html?hpw&rref=health&_r=0

which, among other things, has implications under competition and consumer laws.[4] The question of whether genetically engineered animals are natural is also crucial to inventing (obtaining patent protection) and eating (satisfying food safety regulations) genetically engineered animals, which is the focus of this chapter.

In this chapter we argue that genetically engineered animals present a paradox of nature.[5] More specifically we argue that genetically engineered animals defy simple categorisation and that, depending on the context in which they are situated, genetically engineered animals are at once natural and something other than natural. To support our paradox of nature argument we focus on a particular genetically engineered animal: AquaBounty Technologies' (AquaBounty) genetically engineered Atlantic salmon – branded as AquAdvantage® Salmon[6] – that has been genetically engineered to grow more rapidly than non-genetically engineered Atlantic salmon.[7] In particular, this chapter focuses on the status of this animal under United States patent and food safety law. It is important to point out that in this chapter we do not attempt to address metaphysical, normative or doctrinal questions about what it means to be natural.[8] Rather, we look more broadly at how genetically engineered AquAdvantage Salmon and, by extension, genetically engineered animals compare to other, non-genetically engineered

---

[accessed 29 May 2014]. See also Merchant, G., Cardineau, G. and Redick, T. 2010. *Thwarting Consumer Choice: The Case against Mandatory Labelling for Genetically Modified Foods.* Washington DC: AEI Press.

4   For example, a class action lawsuit was filed in 2012 in the California Northern District Court by Elizabeth Cox, on behalf of herself and others, against the manufacturer of Mission Tortilla chips. Elizabeth Cox alleged that the defendant sold products using false and misleading labelling and advertising because they used the phrase 'All Natural' on products that contained corn from genetically modified seeds: *Cox v. Gruma Corporation*, No. 12-06502, 2013 WL 3828800 (11 July 2013).

5   Throughout this chapter we use 'genetically engineered animals', although others might refer to transgenic or transgene animals or, in Europe, genetically modified animals.

6   The registered trademark symbol (®) will not be used throughout the chapter. Perhaps notably, fish is an important class of case study as world production of farmed fish has now overtaken that of beef: Marshall, M. 2013. Farmed fish overtakes farmed beef for first time. *New Scientist*, 2922. [Online]. Available at: http://www.newscientist.com/article/dn23719-farmed-fish-overtakes-farmed-beef-for-first-time.html#.U3vfqU3HrRY [accessed 20 May 2014].

7   More generally, Michael Carolan has discussed the politics of biotechnology in that '[t]he assemblages that make biotechnology unnatural (and thus patentable) are in fact quickly understated when the topic turns to the regulation of these artefacts.' See also Carolan, M. 2008. From patent law to regulation: the ontological gerrymandering of biotechnology. *Environmental Politics* 17(5), 749, 758.

8   See, for example, Latour, B. 2004. *The Politics of Nature: How to bring the sciences into democracy.* Cambridge, MA: Harvard University Press; Haraway, D. 1991. *Simians, Cyborgs and Women: The reinvention of nature.* New York: Routledge; Armstrong, D. 1983. *What is a Law of Nature?* Cambridge: Cambridge University Press.

Atlantic salmon that are generally accepted as being natural. In this sense, and for the purpose of this chapter, wild Atlantic salmon and Atlantic salmon that are bred or farmed conventionally are treated as 'natural'.

For the most part, in the United States, the patenting of living organisms is legally uncontroversial and there have been numerous patents granted to inventions related to genetically engineered animals.[9] AquAdvantage Salmon is unique because it is not only subject to a United States patent but it is also currently subject to the United States Food and Drug Administration (FDA) regulatory process[10] and has been preliminarily assessed on whether it is safe for human consumption.[11] If approved by the FDA, AquAdvantage Salmon would be the first genetically engineered animal to be commercialised for human consumption.[12] As such AquAdvantage Salmon is a focal point for debates over genetically engineered animals and provides an ideal case study to scrutinise the paradox of nature in United States patent law and food safety regulation. This chapter acknowledges that patent and food safety laws are different regulatory regimes with different social and economic objectives. Further, the patent application and FDA documents were produced by different authors, were intended for different audiences and were prepared at different points in time. Nevertheless, a comparison of how the

---

9 Controversy remains, however, over 'isolated' and 'purified' DNA and DNA sequences: see, *Association for Molecular Pathology v. Myriad Genetics* 569 U.S. (2013), where the United States Supreme Court unanimously concluded that Myriad's claims over a DNA segment were not patentable because Myriad had merely isolated a DNA segment that is found in nature. See also Beauchamp, C. 2013. Patenting nature: A problem of history. *Stanford Technology Law Review*, 16(2), 257–312; Lawson, C. 2014. Patenting DNA sequences after the US myriad decision: New frontiers or just more of the same? *Biotechnology Law Report*, 33(1), 3–16.

10 US Patent No. 5,545,808, *Transgenic Salmonid Fish Expressing Exogenous Salmonid Growth Hormone*, issued on 13 August 1996 to Choy Hew and Garth Fletcher. Later licensed to AquaBounty. To access the documents used in the regulatory process see: United States Food and Drug Administration, *Preliminary Finding of No Significant Impact: AquAdvantage® Salmon* (4 May 2012) and other documents. [Online]. Available at: http://www.fda.gov/AnimalVeterinary/DevelopmentApprovalProcess/GeneticEngineering/GeneticallyEngineeredAnimals/ucm280853.htm [accessed 2 May 2014].

11 See final section below.

12 In November 2013 the Canadian government approved the commercial production of genetically engineered AquAdvantage Salmon eggs on Prince Edward Island. At the time of writing this chapter, however, the Canadian government had not approved the fish for human consumption: see Department of the Environment, *Significant New Activity Notice No. 16528*. [Online]. Available at: http://gazette.gc.ca/rp-pr/p1/2013/2013-11-23/html/notice-avis-eng.html#d106 [accessed 29 May 2014]. The first genetically engineered food crop to be approved for commercialisation by the FDA in 1994 was the Calgene Flavr Savr tomato (a tomato that was genetically engineered to inhibit a gene that produces a protein that makes tomatoes soften). See Bruening, G. and Lyons, J. 2000. The case of the Flavr Savr tomato. *California Agriculture*, 54(4), 6–7 (who outline the 'scientific success, a temporary sales success and then commercial demise' of the Flavr Savr tomato).

two regimes deal with the image of 'natural' is valuable because both the patent and the (possible) FDA approval will attach to the same genetically engineered Atlantic salmon.

The remainder of this chapter is in four parts. The next explains the background to AquAdvantage Salmon technology, its United States patent and its ongoing FDA food safety assessment. The chapter then considers the way in which the underlying technology and the salmon itself are explicitly and implicitly presented in United States patent 5,545,808 and, finally, looks at the relevant, publicly available FDA documents and proceedings that consider whether AquAdvantage Salmon is safe to eat. The analysis shows that for the purpose of gaining patent protection, genetically engineered animals need to have 'markedly different' characteristics from something occurring in nature. The same animals, however, need to be 'substantially equivalent' to non-genetically engineered, natural animals in order to gain food safety approval and be deemed safe to eat. This, as we argue, requires genetically engineered animals to be both natural and something other than natural. In other words, genetically engineered animals are 'different' yet 'equivalent' to non-genetically engineered animals. In conclusion we argue that genetically engineered animals exist in specific legal, social and political contexts and have different meanings depending on these contexts. So, when thinking about whether genetically engineered animals are natural – or toxic, safe, ethical, right and so on – we must be attentive to the different ways of validating and evaluating genetically engineered animals.

Background to AquAdvantage Salmon

In the early 1990s, researchers Garth Fletcher, Peter Davies and Choy Hew at Memorial University of Newfoundland created a gene construct for transgenic Atlantic salmon. Researchers had developed a way of transferring a recombinant deoxyribonucleic acid (rDNA) construct – composed of a promoter from an ocean pout (an eel-like species) antifreeze protein gene and a protein-coding sequence from a Chinook salmon growth hormone – into wild Atlantic salmon fertilised eggs using micro-injection.[13] While Atlantic salmon normally only grow in warm water, from April to September in the Northern hemisphere, the genetically engineered Atlantic salmon produce growth hormones that enable them to grow throughout the year. This means that the genetically engineered Atlantic salmon

---

13   See United States Food and Drug Administration, *Preliminary Finding*, above n 10. To read some of the earlier, related publications see Davies, P., Fletcher, G. and Hew, C. 1989. Fish antifreeze protein genes and their use in transgenic studies, in *Oxford Surveys on Eukaryotic Genes, Volume 6* edited by N. Maclean. Oxford: Oxford University Press, 85–110; Fletcher, G., Shears, M., King, M., Davies, P. and Hew, C. 1988. Evidence for antifreeze protein gene transfer in Atlantic salmon (*Salmo salar*). *Canadian Journal of Fisheries and Aquatic Sciences*, 45(2), 352–7.

reaches 'market size' in about 18 months, compared with at least 30 months for non-genetically engineered Atlantic salmon.[14]

In March 1994 a patent application was filed for the transgenic Atlantic salmon, and on 13 August 1996 Choy Hew and Garth Fletcher were issued United States patent 5,545,808 titled 'Transgenic Salmonoid Fish Expressing Exogenous Salmonoid Growth Hormone'.[15] United States Patent 5,545,808 comprised seven claims related either to genetically engineered Atlantic salmon containing the functioning growth hormone or to methods of increasing the growth rate of Atlantic salmon. While the seven claims are reproduced in full later, the focus of this chapter is on the claims concerning genetically engineered Atlantic salmon.

AquaBounty acquired a license to the genetically engineered Atlantic salmon and the methods of increasing the growth rate of Atlantic salmon from the University of Toronto and Memorial University of Newfoundland.[16] Making use of the licenced technology, AquaBounty spent over a decade and tens of millions of dollars developing the AquAdvantage Salmon.[17] According to AquaBounty, the company will produce AquAdvantage Salmon as triploid (sterile), all-female populations with eyed-eggs as the product for commercial sale and distribution. These eggs would be produced in AquaBounty's facility on Prince Edward Island in Canada[18] and, once produced, sent to a grow-out facility in Panama to reach market size and then harvested for retail sale in the United States as fillets.[19] Under the proposed plan, AquaBounty will not produce or grow AquAdvantage Salmon

---

14   The market size of Atlantic salmon ranges from 2–8kg (4–17lbs): see Veterinary Medicine Advisory Committee, *An overview of Atlantic salmon, its natural history, aquaculture, and genetic engineering.* [Online]. Available at: http://www.fda. gov/AdvisoryCommittees/CommitteesMeetingMaterials/VeterinaryMedicineAdvisory Committee/ucm222635.htm [accessed 2 May 2014].

15   United States Patent No 5,545,808, above n 10.

16   AquaBounty Technologies (originally incorporated in 1991 as A/F Protein) is a biotechnology company based in Massachusetts, United States that is focused on improving productivity in commercial aquaculture. It has the objective of applying biotechnology to ensure the availability of high quality seafood to meet global consumer demand: see AquaBounty Technologies, *The company.* [Online]. Available at: www.aquabounty.com/ company/company [accessed 2 May 2014]. While other companies also licenced the technology they had abandoned their research by 2000: see Reichhardt, T. 2000. Will souped up salmon sink or swim? *Nature,* 406(6791), 10–12.

17   Fox, J. 2010. Transgenic salmon inches toward finish line. *Nature Biotechnology,* 28(11), 1141–2.

18   See Government of Canada, *Canada Gazette,* volume 147 (2013). [Online]. Available at: http://gazette.gc.ca/rp-pr/p1/2013/2013-11-23/html/notice-avis-eng.html#d106 [accessed 29 April 2014].

19   United States Food and Drug Administration, *AquAdvantage® Salmon: Draft Environment Assessment* (4 May 2012). [Online]. Available at: http://www.fda.gov/Animal Veterinary/DevelopmentApprovalProcess/GeneticEngineering/GeneticallyEngineered Animals/ucm280853.htm [accessed 2 May 2014].

in the United States and no live fish will be imported into the United States for processing.[20]

AquAdvantage Salmon have shorter production cycles, increased efficiency of production and, despite the increased speed at which the fish grows, they are the same size at maturity and have the same qualities – such as appearance and taste – as non-genetically engineered Atlantic salmon.[21] AquaBounty claims that these benefits will allow the use of alternative production systems and have substantial environmental and fish health benefits that are not feasible for conventional Atlantic salmon.[22] This, they say, will lower the price of salmon, in part because AquAdvantage Salmon need around 25 per cent less food than conventionally bred fish but also because they reach market size in approximately half the time. Furthermore, AquaBounty suggests that the commercialisation of genetically engineered fish may have a role to play in overcoming depleted fish stocks by providing an alternative to consumption of wild stocks to meet worldwide protein demands.[23] AquaBounty is also developing similar technology to genetically engineer trout and tilapia for faster growth than their non-genetically engineered siblings.[24]

Despite the purported commercial benefits of genetically engineered Atlantic salmon, AquAdvantage Salmon has generated controversy over its potential effects on the environment, human health, culture and trade. An invention such as AquAdvantage Salmon may be patented if it meets the relevant patenting requirements but 'there is no right to *exercise* the invention unless other lawful obligations are satisfied,' such as environmental and food safety regulations.[25] From the time when AquaBounty commenced the process towards gaining regulatory approval for the commercialisation of AquAdvantage Salmon,[26] there has been much discussion, debate and controversy. This was exacerbated when

---

20   FDA, above n 13, 2.

21   AquaBounty Technologies, *Our technology*. [Online]. Available at: http://www. aquabounty.com/technology [accessed 2 May 2014].

22   AquaBounty Technologies, ibid.

23   FDA, above n 19, 6. For a discussion of the state of fish and aquaculture see: FAO, *The state of world fisheries and aquaculture 2012*. [Online]. Available at: http://www.fao. org/docrep/016/i2727e/i2727e.pdf [accessed 2 May 2014].

24   See, for example, United Nations University, *Bioprospecting information resource – Antarctic*. [Online]. Available at: http://www.bioprospector.org/bioprospector/antarctica/entry.action?id=74 [accessed 2 May 2014].

25   Malbon, J., Lawson, C. and Davison, M. 2014. *The WTO Agreement on Trade-Related Aspects of Intellectual Property Rights: A commentary*. Cheltenham: Edward Elgar, 438.

26   The first step in 1995 was to request an investigational exemption for AquAdvantage Salmon under 21 CFR Part 511: see FDA, above n 19, 100. Approvals of this type constitute 'major Federal actions' for which FDA must meet environmental review requirements under the *National Environmental Policy Act* 1969 and FDA's regulations, triggering the requirement to perform an environmental assessment: see, FDA, above n 19, 5.

the FDA released its *Draft Environment Assessment* and *Preliminary Finding of No Significant Impact*.[27]

The approval process for the commercialisation of AquAdvantage Salmon, which is discussed in more detail later, has highlighted questions about the FDA legal and regulatory framework, with some writers suggesting that the FDA processes, which are designed for drugs and not food, are ill-suited to determining the health and environmental safety of using genetically engineered animals as food.[28] Analysts have also raised a number of environmental concerns such as the potential for escape and cross breeding with wild salmon, which may reduce genetic diversity and disease resistance in wild stocks.[29] As the United States *National Environmental Policy Act* 1969 does not require an analysis of the environmental effects in foreign countries,[30] effects on the local environments of the production and grow-out facilities in Canada and Panama were not evaluated in the FDA's *Draft Environmental Assessment*.[31] The FDA concluded that the production of AquAdvantage Salmon (outside of the United States) 'will not result in significant effects on the quality of the human environment in the United States'.[32]

There have also been concerns raised over the unknown and unintended consequences of consuming genetically engineered food.[33] Some commentators, for example, have argued that AquAdvantage Salmon has higher levels of endogenous

---

27   FDA, ibid.; FDA, above n 13.

28   See, for example, Wilinska, K. 2012. AquAdvantage is not real advantage: European biotechnology regulations and the United States' September 2010 FDA Review of Genetically Modified Salmon. *Minnesota Journal of International Law*, 21(1), 145–76. For an alternative view see Noah, L. 2013. Whatever happened to the "Frankenfish"? The FDA's foot dragging on transgenic salmon. *Maine Law Review*, 65(2), 232–51 (who considers the majority of the criticism of genetically engineered salmon to be misplaced and largely political).

29   See, for example, Bernhard Bentsen, H. and Thodesen, J. 2005. Genetic interactions between farmed and wild fish, with examples from the Atlantic salmon case in Norway, in *Selection and Breeding Programs in Aquaculture* edited by T. Gjedrem. New York: Springer, 319. AquAdvantage claims to have addressed environmental concerns through the use of physical, biological (production of sterile female only populations) and geographical/geophysical (land-based facilities) forms of containment. See FDA, above n 13, 3. See also Geiger, K. 2010. Genetically modified salmon safe to eat, FDA Report Says', *LA Times*, 4 September 2010. [Online]. Available at: http://articles.latimes.com/2010/sep/03/nation/la-na-fda-salmon-20100904 [accessed 2 May 2014]. But see, Noah, ibid.

30   *National Environmental Policy Act* 1969 42 U.S.C. § 4321 *et seq*. See FDA, above n 13, 3.

31   Only the effects on the United States of exposure pathways from these facilities were evaluated: see FDA, above n 19, 2.

32   FDA, ibid., 3. The FDA went on to say that approval of AquaBounty's application will have no effect on endangered species or critical habitat within the United States, FDA, ibid., 11. See also FDA, above n 13, 5.

33   Other concerns include the disruption to the international salmon trade arising from tensions between the United States and its trading partners, such as the European

allergenic potency that could be harmful to humans and that it has the potential to act as a vector for human pathogens.[34] In its preliminary assessment, however, the FDA concluded that the AquAdvantage Salmon is 'as safe as food from conventional salmon and that there is a reasonable certainty of no harm from consumption of food from triploid AquAdvantage Salmon'.[35] In response to the FDA's preliminary findings, over 1.8 million people expressed their opposition to AquAdvantage Salmon in 2013.[36] Furthermore, on 3 March 2014, two of the largest retail grocers in the United States, Kroger and Safeway, agreed not to sell genetically engineered salmon.[37]

The rest of this chapter examines how the genetically engineered Atlantic salmon and AquAdvantage Salmon were presented to the United States Patent and Trademark Office (USPTO) and FDA respectively. United States patent law and food safety regulations draw out a paradox of nature. This paradox of nature is examined in the next two sections of this chapter. First, we consider the need for an invention under United States patent law to have 'markedly different' characteristics from something occurring in nature. Second, we consider the requirement that genetically engineered animals be 'substantially equivalent' to comparative, non-genetically engineered animals for the purpose of food safety; for this requirement the FDA considers, among other things, whether the essential nature of the animal has changed and whether any genetic variations are the result of natural variation and consequently safe to eat.

Union, who are largely opposed to genetically engineered animals entering the food supply: see Wilinska, above n 28, 158–70.

34   See generally, Fisheries and Oceans Canada, *Summary of the Environmental and Indirect Human Health Risk Assessment of AquAdvantage Salmon*, Canadian Science Advisory Secretariat, Science Response 2013/023. [Online]. Available at: http://www.dfo-mpo.gc.ca/csas-sccs/Publications/ScR-RS/2013/2013_023-eng.pdf [accessed 2 May 2014]. The report stated that 'due to a lack of data, no conclusion can be drawn as to whether [AquAdvantage Salmon] would have an increased capacity to act as a reservoir for the transmission of disease agents to humans' (at 13).

35   FDA, above n 13, 3.

36   Center for Food Safety, Nearly 2 million people tell FDA not to approve GE salmon. [Online]. Available at: http://www.centerforfoodsafety.org/press-releases/2151/nearly-2-million-people-tell-fda-not-to-approve-ge-salmon# [accessed 2 May 2014]. The effort was driven by a broad coalition organised over three years ago by the Center for Food Safety and consisting of public interest, consumer, environmental and animal protection groups, along with commercial and recreational fisheries associations and food businesses and retailers. Due to public and political pressure the FDA extended the comment period on the *Draft Environmental Assessment* and *Preliminary Finding of No Significant Impact* for 60 days until 26 April 2013: United States Food and Drug Administration, *FDA Extends Comment Period on AquaAdvantage Salmon Documents*, 13 February 2013. [Online]. Available at: http://www.fda.gov/AnimalVeterinary/NewsEvents/CVMUpdates/ucm339270.htm [accessed 2 May 2014].

37   Friends of the Earth, *Kroger, Safeway join trend away from GMO food*. [Online]. Available at: http://www.foe.org/news/archives/2014-03-kroger-safeway-join-trend-away-from-gmo-food [accessed 2 May 2014].

Genetically Engineered Animals That Are 'Markedly Different' from Something Occurring in Nature Are Patentable

Under United States patent law, an invention may be patentable if it is patent-eligible subject matter such as a process, machine, manufacture or composition of matter,[38] new,[39] useful,[40] non-obvious[41] and sufficiently disclosed/described.[42] The patentable subject matter requirement has received intense scrutiny in the context of biotechnological patent claims because, in some ways, biotechnological innovations have challenged the product of nature doctrine, a long-standing limit on patentable subject matter.[43] Being one of the most published areas of patent law, not much can be written about the product of nature doctrine that has not already been written.[44] Nevertheless, when considering the patenting of genetically engineered animals and the paradox of nature in inventing and eating genetically engineered animals it is necessary to briefly outline the product of nature doctrine.

The product of nature doctrine can be traced back to the nineteenth century decision of *Ex parte Latimer* and the statement in American law that to patent 'trees of the forest and the plants of the earth' would be 'unreasonable and impossible'.[45] A significant case in relation to the product of nature doctrine is *Funk Brothers Seed Co v. Kalo Inoculant Co* where the United States Supreme Court held a patent to be invalid 'for want of invention' because the claimed invention had been discovered and not invented. In *Funk Brothers*, the United States Supreme Court

---

38   United States *Patent Act 1952* 35 USC §101.

39   United States *Patent Act 1952* 35 USC §102.

40   United States *Patent Act 1952* 35 USC §101.

41   United States *Patent Act 1952* 35 USC §103.

42   United States *Patent Act 1952* 35 USC §112.

43   Beauchamp argues that leading cases in the area often blended consideration of patentable subject matter with the other patentability requirement when applying the doctrine. However, the United States Supreme Court in *Mayo Collaborative Services v. Prometheus Laboratories, Inc* 132 S. Ct. 1289, 1303–04 (2012) has declared that the product of nature cases 'rest their holdings on section 101, not later sections'. Beauchamp, above n 9, 265.

44   See for example, Beauchamp, ibid.; Jackson, J. 2010. Something like the sun: Why even 'isolated and purified' genes are still products of nature. *Texas Law Review*, 89(6), 1453–89; Conley, J. and Makowski, M. 2003. Back to the future: Rethinking the product of nature doctrine as a barrier to biotechnology patents (Part I), *Journal of the Patent and Trademark Office Society*, 85(3), 301–34; Conley, J. Gene patents and the product of nature doctrine. *Chicago-Kent Law Review*, 84(1), 109–32.

45   *Ex Parte Latimer* 1889 Dec. Comm'r Pat. 123 [1989] 125–6. The extracted pine fibre in question was 'a natural product and can no more be the subject of a patent in its natural state when freed from its surroundings than wheat which has been cut by ... some new method of reaping can be patented as wheat cut by such a process' (at 127). See Beauchamp, above n 9, 273 (who points out that this was a decision of the Patent Commissioner, not a court).

reasoned that the claimed invention failed to 'disclose an invention or discovery within the meaning of the patent statutes'.[46] In so doing, the Supreme Court laid out the product of nature doctrine stating:

> ... patents cannot issue for the discovery of the phenomena of nature ... The qualities of these bacteria, like the heat of the sun, electricity, and or the qualities of metals, are part of the storehouse of knowledge of all men ... He who discovers a hitherto unknown phenomenon of nature has no claim to a monopoly of it which the law recognises. If there is to be invention from such a discovery, it must come from the application of the law of nature to a new and useful end.[47]

Perhaps the most often cited and discussed legal decision in the area of genetically engineered animals is the 1980 decision of the United States Supreme Court in *Diamond v. Chakrabarty*, a case involving a patent application for a genetically engineered bacterium capable of 'digesting' multiple components of crude oil.[48] In *Diamond v. Chakrabarty*, the Supreme Court ruled that living organisms may be patented. The claimed bacterium was not found in nature nor was its activity exhibited in any naturally occurring bacteria. The Court pointed out that the bacterium is 'a non-naturally occurring manufacture or composition of matter – a product of human ingenuity "having a distinctive name, character and use"'.[49] Furthermore, the Court emphasised that the claimed bacterium has '*markedly different* characteristics from any found in nature and ... is not nature's handiwork, but his own'.[50] Therefore, the critical reason for holding that the bacteria were patentable was that it did not exist in nature, it could not be created in nature and it had a property not found in nature. In its often-quoted opinion, the Supreme Court cited a 1952 Senate Report that said that 'anything under the sun made by man' is patentable as long as the conditions of the title are fulfilled.[51]

Following the reasoning in *Chakrabarty*, the United States Board of Patent Appeals and Interferences determined that animals are patentable subject matter in the United States. In *Ex parte Allen*, for example, the Board decided that a polyploid Pacific coast oyster could have been the proper subject of a patent if all

---

46   *Funk Brothers Seed Co. v. Kalo Inoculant Co.* 333 U.S. 127 [1948] 138.
47   *Funk Brothers Seed Co. v. Kalo Inoculant Co.* 333 U.S. 127 [1948] 130.
48   *Diamond v. Chakrabarty*, 447 U.S. 303 (1980).
49   Ibid., 309–10 (quoting *Hartranft v. Wiegmann*, 121 U.S. 609, 615 (1887).
50   Ibid., 310 (emphasis added).
51   Ibid., 309. This quote is often taken out of context, however, and does not indicate the 'expansive notion of patentable subject matter but a limit on its reach': Feldman, R. 2012. *Rethinking patent law*. Boston: Harvard University Press.

the criteria for patentability were satisfied.[52] On 21 April 1987, the Commissioner of Patents and Trademarks issued a notice, *Animals – Patentability, 1077 O.G. 24*, that made it clear that the USPTO considered non-naturally occurring, nonhuman multicellular living organisms, including animals, to be patentable subject matter. In 1988 the USPTO issued its first patent for a transgenic animal for the Harvard Oncomouse mouse, a transgenic mouse with a predictable tendency to develop breast cancer.[53] Since the Oncomouse patent was granted in 1988, hundreds of patents relating to genetically engineered animals have been issued in the United States including those over pigs, cows, chickens, monkeys, rabbits and sheep.[54]

In summary, then, the test set down in the United States for patentable subject matter in this area is whether the invention is the result of human intervention and has 'markedly different' characteristics from something occurring in nature. The relevant distinction, therefore, is not between living and non-living things but between products of nature and human-made inventions. While United States patent 5,545,808 does not claim the DNA sequence, but instead claims the entire genetically engineered salmon that contains the DNA construct, it is worth pointing out that many of the legal issues with patenting genes and genetically engineered animals are associated with claims to DNA and DNA sequences.[55] The materiality of the differences between claimed DNA sequences and their natural counterparts turns on whether they are claimed as chemical compositions (artificial/synthetic/unnatural) or as descriptive sequence information (natural), the latter of which is

---

52   *Ex Parte Allen*, 2 U.S.P.Q.2d 1475 (Federal Circuit, 1987). The application was denied, however, because of the obviousness of the claimed invention. Also see *American Fruit Growers Inc v. Brogdex Co* 283 U.S. 1, 13 (1931) ('There must be a transformation; a new and different article must emerge having a distinctive name, character, or use').

53   United States Patent No. 4,736,866, *Transgenic non-human mammals* (granted 12 April 1988).

54   See for example, Crane, A. 2009. Of mice and men: Patentability of genetic material and the protection of intellectual property rights. *Dalhousie Journal of Legal Studies*, 18(1), 93–117; Woessner, W. 2001. The evolution of patents on life: Transgenic animals, clones and stem cells. *Journal of the Patent and Trademark Office Society*, 83(11), 830–44; Lane, M. 1991. Patenting life: Responses of patent offices in the U.S. and abroad. *Jurimetrics*, 32(1), 89–100; Montgomery, M. 1990–1991. Building a better mouse – and patenting it. Altering the patent law to accommodate multicellular organisms. *Case Western Law Review*, 41(1), 231–65.

55   A doctrine that has been conceptualised as either an exception to, or loop hole for, the product of nature doctrine is the idea that purification or isolation alone can distinguish a claimed product from a naturally occurring counterpart: See for example Beauchamp, above n 9; Lawson, above n 9, 23; Conley, J. and Makowski, R. 2004. Rethinking the product of nature doctrine as a barrier to biotechnology patents in the United States – and perhaps Europe as well. *Information and Communications Technology Law*, 13(1), 3–40.

not patent eligible.[56] Patent lawyers, therefore, often treat the product of nature requirement, as a technical, claim-drafting problem.[57]

Central to establishing that genetically engineered animals are not a product of nature is the patent application. This document is used to convince the USPTO that the claimed invention is *inter alia* proper (patentable) subject matter which, as we have seen, depends on the invention having 'markedly different' characteristics from any found in nature and being the result of human intervention.[58] A patent application has a number of parts. In the United States it consists of a title, an abstract, claims and a description of the invention.[59] While there is some variation in the headings used in patents, they must include bibliographical data referring to other related patents, drawings of the invention and a specification.[60] The specification is usually divided into a number of sections including the field of the invention, background of the invention, a detailed description (the preferred embodiment) that sets out 'the manner and process of making and using it, in such full, clear, concise, and exact terms as to enable any person skilled in the art ... [to] make and use the same', and a brief description of any drawings included in the application.[61] The patent application is sent to the USPTO where it is assigned an examiner who then determines whether the application meets the statutory requirements.

Although a patent application sets out a particular factual situation and is used by the patent examiner to determine whether the claimed invention is eligible for protection, it also has broader, less doctrinal dimensions, including influencing a patent examiner to form a particular judgement through the use of language and imagery. In an article titled 'From Discovery to Invention: The Writing and Rewriting of Two Patents' David Meyers points out that, despite the similarities of journal articles and patent applications – such as that both typically have a title, an author, an abstract, some claims, a narrative and references, they differ in their presentation of claims, their relation to other texts and their narrative construction of future action.[62]

---

56   The USPTO has said that '[w]hile descriptive sequence information alone is not patentable subject matter, a new and useful purified and isolated DNA compound described by the sequence is eligible for patenting, subject to satisfying the other criteria for patentability:' United States Patent and Trademark Office, *Utility Examination and Guidelines*, 66 Fed. Reg. 1092, 1093 (2001).

57   Eisenberg, R. 2000. Re-examining the role of patents in appropriating the value of DNA sequences. *Emory Law Journal*, 49(3), 783–800, 785.

58   Patentable subject matter is defined as 'any new and useful process, machine, manufacture or composition of matter, or any new and useful improvement thereof': 35 U.S.C. § 101. As we have noted, the claimed invention also needs to be novel (35 U.S.C. § 102), non-obvious (35 U.S.C. § 103) and have utility (35 U.S.C. § 101).

59   Meyers, G. 1995. From discovery to invention: The writing and rewriting of two patents. *Social Studies of Science*, 25(1), 57–105.

60   US Code Title 35, Part II, Chapter 11.

61   35 U.S.C. § 112.

62   Meyers, above n 59, 58.

This, according to Meyers, is significant because it highlights the different types of professional expertise embodied in patent applications and scientific papers.[63] Other academics have noted the rhetorical and stylistic aspects of patents, including Charles Bazerman (who argues that patents were used, not only to obtain a right over the invention, but to stabilise a shifting world, to create order in an uncertain world and to give the new technology meaning and value)[64] and Geof Bowker (who argues that patents must be seen in relation to a legal strategy that involves changes in the market and in technical practice, as well as the more narrowly defined inventions in the patent).[65]

Importantly, this chapter does not question the validity of United States patent 5,545,808.[66] Rather, it identifies the legal and strategic use of information and material presented in the patent application so as to distinguish genetically engineered animals from animals existing in nature. In terms of AquAdvantage Salmon we look at three particular aspects of the patent application: the images, descriptions and claims. One of the ways in which the genetically engineered Atlantic salmon is presented in United States patent 5,545,808 is through the inclusion of images. The application has seven drawings in total that highlight the scientific nature of the invention and includes polymerase chain reaction (PCR) analysis, ocean pout antifreeze protein (op-AFT) promoter activity in chloramphenicol acetyltransferase (CAT) assay and the DNA sequence of the transgenic salmon. Furthermore, out of the seven drawings, there is one image of fish that compares the size of the claimed genetically engineered Atlantic salmon with non-genetically engineered salmon at the same age. The included image is of such a resolution and quality that, other than comparing the sizes of the Atlantic salmon at the same point in time, the images do not particularly look like, or identify, an Atlantic salmon.

Another way that the genetically engineered Atlantic salmon is distinguished from nature in United States patent 5,545,808 is in the way the invention is described. In patent law, the description has a technical purpose and is used to specify the claimed invention and how it differs from previous patents and

---

63   See also Swanson, K. 2011. Authoring an invention: Patent production in the nineteenth century United States' in *Making and Unmaking Intellectual Property: Creative production in legal and cultural perspective* edited by M. Biagioli, P. Jaszi and M. Woodmansee. Chicago: University of Chicago Press, 41–54 (arguing that a patent application is a bureaucratic text).

64   Bazerman, C. 1999. *The Languages of Edison's Light: Rhetorical agency in the material production of technology.* Cambridge MA: MIT Press, 1–2. See also Bazerman, C. 1993. Patent realities: Legally stabilized texts and market indeterminacies, in *The Narrative Construction of the Anxious Object* edited by J. Hultberg. Goteborg: University of Goteborg, 5–12.

65   Bowker, G. 1992. What's in a patent? in *Constructing Stable Technology.* W. Bijker and J. Law (eds), Cambridge MA: MIT Press, 53–74.

66   United States patent 5,545,808 does not claim an isolated gene (compare with the *Myriad* example above n 9).

technology. It also has the purpose of distinguishing the claimed invention from naturally occurring Atlantic salmon and highlighting the human intervention required to 'invent' the genetically engineered Atlantic salmon. United States patent 5,545,808 is heavy with scientific literature and terminology that points out the problems of earlier scientific research and how the claimed invention advances existing science and technology. The invention is described as '[a] chimeric gene, pOP-GHe ... constructed by using antifreeze promoter linked to the chinook salmon GH cDNA clone'.[67] And, with sub-headings like 'Experimental Procedures', 'Plasmid Construction' and 'Test Results of the Above Procedures' the description resembles a scientific report. The patent text contains over 35 scientific references from journals including *DNA, Biotechniques, Theoretical Applied Genetics, Fish Physiology* and *Biochemistry and Biophysics*, setting out the difficulties of developing genetically engineered fish. It also lists the DNA sequence in the Atlantic salmon:

> The 4kb insert in pOP-GHe was excised by Eco RI digestion and dissolved in saline buffer at a concentration of 3.mu.g/ml. Approximately (2–3 nl, 10.sup.6 copy) of the DNA insert was injected through the micropyle into a fertilized, nonactivated salmon egg cytoplasm (Fletcher et al. 1988) in accordance with the procedure of the aforementioned U.S. patent application Ser. No. 278,463. Approximately 500 eggs were injected. The survival rate was 80% as compared to the noninjected control.[68]

Yet another, and perhaps the most important, way in which the genetically engineered Atlantic salmon is presented in the patent application is in the claims. The claims define the invention and set out the rights that will be possessed by the patentee. They also convince patent examiners, lawyers and judges that the claimed genetically engineered animal is patentable subject matter and are thus used to help distinguish the claimed invention from nature. Indeed, before the USPTO can examine novelty, non-obviousness and description,[69] it first must determine the threshold issue of whether the claimed invention is patentable (proper) subject matter.[70]

The AquAdvantage patent makes seven claims in total, some of which are over genetically engineered Atlantic salmon (claims 4, 5, 6 and 7) and others over the methods of increasing the growth rate of Atlantic salmon (claims 1, 2 and 3). The manufacture claims are:

---

67   United States Patent 5,545,808, 4–5.

68   United States Patent 5,545,808, 7–8.

69   35 U.S.C. § 102, 103 and 112 (2006).

70   35 U.S.C. § 101 (2006). United States Patent and Trade Mark Office, 2012. *Manual of patent examining procedure* § 2105 (eighth edition, 2001, revised 2012). [Online]. Available at: http://www.uspto.gov/web/offices/pac/mpep/ [accessed 2 May 2014].

A transgenic salmonid fish containing in its germline a salmonid growth hormone gene operably linked to a type 3 antifreeze protein promoter wherein said salmonid fish expresses said growth hormone gene at levels which increase the rate of its growth at least four times that of a salmonid fish lacking said growth hormone gene operably linked to said antifreeze protein promoter [claim 4].

A transgenic salmonid fish of claim 4 wherein said antifreeze protein promoter is from ocean pout and said growth hormone gene is the endogenous hormone gene [claim 5].

A transgenic salmonid fish of claim 4 wherein said salmonid fish is selected from the group consisting of Atlantic salmon and Chinook salmon [claim 6].

A transgenic salmonid fish of claim 5 wherein said salmonid fish is Atlantic salmon [claim 7].

The process claims are:

A method of increasing the growth rate of a salmonid fish comprising the steps of:

a) introducing into the germ line of a salmonid fish a gene encoding a salmonid growth hormone operably linked to a type 3 antifreeze protein promoter; and,

b) culturing said salmonid fish under conditions wherein said salmonid fish expresses said growth hormone gene at levels which increase the rate of its growth at least four times that of a salmonid fish lacking said growth hormone gene operably linked to said antifreeze protein promoter [claim 1].

A method of claim 1 wherein said antifreeze protein promoter is from ocean pout and said growth hormone gene is the endogenous growth hormone gene [claim 2].

A method of claim 1 wherein said salmonid fish is selected from the group consisting of Atlantic salmon and Chinook salmon [claim 3].

Assessing claims over the genetically engineered Atlantic salmon (the 'manufacture' claims) required the claimed invention to be distinguished from nature.[71] In *Diamond v. Chakrabarty*, the court established that the term 'manufacture' is expansive and referred to its dictionary definition to conclude that manufacture means the production of articles for use from raw materials, prepared by giving these materials new forms, qualities, properties, or combinations

---

71   Under the United States Patents Act there are four categories of patentable subject matter: process, machine, manufacture and composition of matter: 35 U.S.C. § 101.

whether by hand labour or by machinery.'[72] So under United States patent law, AquaBounty's manufacture claims satisfied the patent examiner that the claimed genetically engineered animal had 'markedly different' characteristics from any found in nature and 'new qualities' to those which exists in nature.[73]

The images, descriptions and claims included in United States patent 5,545,808 were used to satisfy the patent examiner that the genetically engineered Atlantic salmon was patentable subject matter. As we have shown, there was no clear image of an Atlantic salmon, and the images, descriptions and claims all reflect the scientific, interventionist nature of the genetically engineered salmon. Moreover, the images helped convince the USPTO that Atlantic salmon containing the added growth hormone and promoter only exists because of the result of the researchers' knowledge, hard-work and intervention. Finally, the claims were also written in such a way as to convince the patent examiner that the claimed Atlantic salmon was 'markedly different', had 'new qualities' and had been transformed to a different state to Atlantic salmon already in existence. This, as we will see, contrasts to the way in which genetically engineered salmon was presented to the FDA.

Genetically Engineered Animals that are 'Substantially Equivalent' to Non-Genetically Engineered Animals are Safe to Eat

Since 1986, the FDA has asserted jurisdiction over genetically engineered animals, including fish, on the basis that the transgene and any expressed proteins affect the 'structure and function' of the receiving animal in a similar way to veterinary drug formulations.[74] The FDA has the jurisdiction to regulate genetically engineered animals under the 'new animal drug' provisions of the *Federal Food, Drug and Cosmetic Act* 1938 (*FD&C Act*).[75] The definition of a drug in the *FD&C Act* includes 'articles (other than food) intended to affect the structure or any function of the body of man or other animals'.[76] The definition of 'new animal drug' includes a drug intended for use in animals that is not generally recognised as safe and effective for use under the conditions prescribed, recommended, or suggested in the drug's labelling and that has not been used to a material extent or for a material time.[77] The recombinant DNA (rDNA) used to make the AquAdvantage Salmon,

---

72   *Diamond v. Chakrabarty* 447 U.S. 303, 308 (1980). Recently, this was applied by the United States Supreme Court to DNA sequences (as 'compositions of matter') in *Association for Molecular Pathology v. Myriad Genetics* 569 U.S. (2013). For a discussion see Lawson, above n 9.

73   *Diamond v. Chakrabarty* 447 U.S. 303, 308, 310 (1980).

74   United States Office of Science and Technology Policy, *Coordinated Framework for the Regulation of Biotechnology* 51 Fed Reg (26 June 1986).

75   *Federal Food, Drug and Cosmetic Act 1938* 21 U.S.C. 321 *et seq.*

76   *Federal Food, Drug and Cosmetic Act 1938* 21 U.S.C. 321(g).

77   *Federal Food, Drug and Cosmetic Act 1938* 21 U.S.C. 321(v).

therefore, meets the definition of a 'new animal drug' and must be approved by FDA prior to commercialisation.

The *FD&C Act* and Title 21 ('Food and Drugs') of the *Code of Federal Regulations* (*CFR*) sets out the information that must be submitted to the FDA in order to have genetically engineered animals approved for commercialisation.[78] Those seeking FDA approval for a genetically engineered animal must submit a range of data and information including full reports of investigations to show whether the genetically engineered animal is safe and effective for use, a full statement of the composition of the genetically engineered animal, and information on the safety and effectiveness of the genetically engineered animal for its intended use. When deciding whether to approve genetically engineered animals the FDA or, more accurately, the Center for Veterinary Medicine (CVM), takes advice from the Veterinary Medicine Advisory Committee (VMAC) on whether the gene construct is safe in the Atlantic salmon, effective, safe for consumers and unlikely to escape or cause problems for wild Atlantic salmon. On 19 and 20 September 2010, the VMAC held a public meeting to discuss the FDA's assessment of AquAdvantage Salmon.[79]

To assist developers of genetically engineered animals and to clarify the requirements under the *FD&C Act* and *CFR*, in 2009 the FDA published *Guidance for Industry 187: Regulation of Genetically Engineered Animals Containing Heritable DNA Constructs* ('*Guidance 187*').[80] Whilst *Guidance 187* is non-binding, and alternative approaches are allowed, it interprets and sets out the rules to apply to genetically engineered animals, particularly those genetically engineered animals containing heritable rDNA constructs that are compliant with the relevant legal and regulatory frameworks. *Guidance 187* sets out a number of steps for assessing the potential hazards and likelihood of harm that are 'cumulative and risk-based'.[81] The seven steps are: (1) product identification; (2) molecular characterisation of the construct; (3) molecular characterisation of the GE animal lineage; (4) phenotypic characterisation of the GE animal; (5) genotypic and phenotypic durability assessment; (6) food/feed safety and environmental assessments and (7) effectiveness/claim validation.[82]

---

78   Section 512(b)(i) and section 514.1. Further explained in 21 CFR 514.1 (revised as of 1 April 2013) and Guidance 187.

79   The Food and Drug Administration documents on *Genetically engineered salmon*. [Online] Available at: http://www.fda.gov/AnimalVeterinary/Development ApprovalProcess/GeneticEngineering/GeneticallyEngineeredAnimals/ucm280853.htm [accessed 2 May 2014].

80   Food and Drug Administration, *Guidance for Industry 187: Regulation of Genetically engineered animals containing heritable DNA constructs* ((2009) revised in 2011). [Online]. Available at http://www.fda.gov/downloads/animalveterinary/guidance complianceenforcement/guidanceforindustry/ucm113903.pdf [accessed 2 May 2014].

81   Ibid., 20ff.

82   For a summary of these steps see Veterinary Medicine Advisory Committee, FDA Centre for Veterinary Medicine, *Briefing packet: AquAdvantage Salmon* (2010) 3–6.

The decision by the CVM is largely based on the information and materials supplied by the applicant, in this case AquaBounty. This means that insights about the FDA's food safety assessment can be drawn from FDA documents, including the FDA's *Briefing Packet* that provides a summary of the FDA's risk-based process for determining the safety and effectiveness of the AquAdvantage Salmon and describes the scientific design and test data of the AquAdvantage Salmon.[83] The *Briefing Packet* contains a summary of the information on AquAdvantage Salmon's environmental and health risk assessment, based on controlled and non-controlled studies on AquAdvantage Salmon and non-genetically engineered Atlantic salmon, historical data and other scientific studies on AquAdvantage Salmon.

What is interesting about the information in the *Briefing Packet*, particularly in the context of this chapter, is the way in which AquAdvantage Salmon is presented. In attempting to convince the FDA that AquAdvantage Salmon is safe to eat, AquaBounty presents its genetically engineered Atlantic salmon in a way that is not 'markedly different' from 'natural', non-genetically engineered Atlantic salmon. Perhaps most broadly (and in contrast to the patent application) AquaBounty describes its product in a way that downplays the human intervention required. In fact, in the *Briefing Packet* there is no use of the term 'technology' or 'invention' in relation to the product.[84] Instead the term 'construct' is used which is a more neutral term for using existing (nature's) building blocks to build or make something. They indicate that there is nothing novel about the AquAdvantage Salmon, and that AquAdvantage Salmon, and the 'constructs' required to produce it, are nothing more than standard, typical and routine. More specifically, AquAdvantage Salmon was described as having 'a standard plasmid backbone, transcriptional regulatory elements derived from ocean pout, a fish protein growth regulator (Chinook salmon growth hormone), and synthetic primers'[85] and the assembly of the growth hormone construct was described as 'typical for the time that the construct was assembled and utilized routine rDNA procedures'.[86] The description of the technology as standard, typical and routine is evidence that it was important for the food safety aspect of AquAdvantage's application to draw genetically engineered animals, and the technology used in their creation, as close to nature or as natural as possible.

[Online]. Available at http://www.fda.gov/downloads/AdvisoryCommittees/Committees MeetingMaterials/VeterinaryMedicineAdvisoryCommittee/UCM224762.pdf [accessed 2 May 2014] ('FDA Briefing Packet').

83   Ibid. For a list of the materials that the FDA compiles see 21 C.F.R. § 14.33(f) (2012).

84   The 22 times that 'technology' 'technological' or 'technologies' appears in the *Briefing Packet*, ibid., relate to: (a) the FDA review process generally; (b) names (eg AquaBounty Technologies and government divisions/groups) and (c) other literature references. It is interesting to note that AquaBounty Technologies has been abbreviated in the document to ABT which disguises its involvement in technology.

85   FDA, above n 82, 11.

86   Ibid.,12.

The food safety assessment is set out in 21 *CFR* 514.1(b)(8) and considers whether there are any differences between food from genetically engineered animals and similar, non-genetically engineered animals. In the case of AquAdvantage Salmon, therefore, the question is: are there any differences between food derived from AquAdvantage Salmon and other Atlantic salmon? In order to determine whether genetically engineered animals are safe to eat the FDA conducts a comparative analysis.[87] Specifically, the FDA looks at changes in the physiology and composition of edible tissues of the genetically engineered animals for the purpose of assessing toxicology and allergenic potential. The comparative approach acknowledges that non-genetically engineered animals contain some toxins and that these are generally safe for human consumption. Wild and farmed Atlantic salmon, for example, contain polychlorinated biphenyls (PCBs), dieldrin, toxaphene, dioxins and polybrominated diphenyl ethers[88] that have adverse effects on human health[89] beyond tolerable daily intake levels.[90]

The test or standard used to determine whether genetically engineered animals are safe to eat in the FDA's comparative analysis is 'substantial equivalence':[91] if a genetically engineered animal is substantially equivalent to its 'natural' antecedent it is assumed that the genetically engineered animal poses no new or different health risks.[92] Put more simply, if genetically engineered AquAdvantage Salmon looks and tastes like non-genetically engineered Atlantic salmon then it is substantially equivalent to Atlantic salmon and is therefore safe to eat. According to the FDA:

---

87   See, for example, FDA, above n 82; FDA, above n 80.

88   Most studies show that contaminants in farmed animals are higher than those sourced from the wild. See, for example, Hites, R., Foran, J., Schwager, S., Knuth, B., Hamilton, M. and Carpenter, D. 2004. Global assessment of polybrominated diphenyl ethers in farmed and wild salmon. *Environmental Science and Technology*, 38(19), 4945–9.

89   For example, carcinogenicity, immunotoxicity and endocrine disruption effects. See, Foran, J., Good, D., Carpenter, D., Hamilton, M., Knuth, B. and Schwager, S. 2005. Quantitative analysis of the benefits and risks of consuming farmed and wild salmon. *The Journal of Nutrition*, 135(11), 2639–43.

90   See, for example, a discussion on the World Health Organisation's levels of tolerable daily intake of dioxins in Foran, J., Carpenter, D., Hamilton, M., Knuth, B. and Schwager, S. 2005. Risk-based consumption advice for farmed Atlantic and wild Pacific salmon contaminated with dioxins and dioxin-like compounds. *Environmental Health Perspectives*, 113(5), 552–6, 553.

91   Also used in Canada, Japan, the FAO and WHO.

92   See, for example, Grossman, M. 2010. Genetically modified crops and food in the United States: The Federal regulatory framework, state measures, and liability in tort in *The Regulation of Genetically Modified Organisms: comparative approach*. Luc Bodiguel and Michael Cardwell (eds). Oxford: Oxford University Press, 301, 312; Pollack, M. and Shaffer, G. 2009. *When Cooperation Fails: The international law and politics of genetically modified foods*. Oxford: Oxford University Press.

... if the expression product(s) is shown to be safe, and the composition of edible tissues from the GE animal is shown to be as safe as those from animals of the same or comparable type that are commonly and safely consumed, then we expect to view this as evidence that food and feed derived from the GE animal is safe (i.e., there is a reasonable certainty of no harm from consumption of the food or feed).[93]

The determination of whether genetically engineered animals are safe to eat is broken down by the FDA into two specific questions:[94] First, is AquAdvantage Salmon Atlantic salmon? Second, are there any changes in AquAdvantage Salmon compared with non-genetically engineered Atlantic salmon that pose food safety risks? The way in which the AquAdvantage Salmon is presented and assessed for each of these questions will be considered separately below.

The first question that the FDA considers to determine food safety is whether AquAdvantage Salmon is Atlantic salmon. Here, the question is whether the 'essential nature of the salmon' has changed as a result of the introduction of the AquAdvantage construct.[95] This is done with reference to the FDA's Regulatory Fish Encyclopaedia (RFE), a searchable compendium of approximately 1,700 species of finfish and shellfish that is used to identify fish species. The RFE was developed by FDA scientists at the Seafood Products Research Center (Seattle District) and the Center for Food Safety and Applied Nutrition to help federal, state and local officials and purchasers of seafood identify species substitution and economic deception in the marketplace. For each species of finfish and shellfish, the RFE includes acceptable market name(s), the scientific common name and the scientific name as well as vernacular names for cross reference.

While the RFE was originally developed to stop economic fraud and the mislabelling of fish, the FDA uses the RFE for its food safety assessment. One of the ways in which the RFE is used is to provide a photograph and details about Atlantic salmon, which can be compared to the AquAdvantage Salmon in order to see if it looks like Atlantic salmon.[96] The RFE also includes methods for comparing fish such as biochemical patterns. To make the comparison, various samples of genetically

---

93   FDA, above n 80, 24. See also *Joint FAO/WHO expert consultation on biotechnology and food safety* (Rome, Italy, 30 September to 4 October 1996), 4. [Online]. Available at: ftp://ftp.fao.org/es/esn/food/biotechnology.pdf [accessed 2 May 2014].

94   FDA, above n 82 section VII.

95   Ibid.

96   While not included in the *Briefing Packet*, ibid., an image of an Atlantic Salmon from the FDA's Regulatory Fish Encyclopaedia was shown at the VMAC meeting on 20 September 2010. See Greenlees, K. and Jones, K. *Food Safety Assessment*. [Online]. The transcript is available at: http://www.fda.gov/downloads/AdvisoryCommittees/Committees MeetingMaterials/VeterinaryMedicineAdvisoryCommittee/UCM230471.pdf [accessed 2 May 2014]. The slides, including the images, are available at: http://www.fda.gov/ downloads/AdvisoryCommittees/CommitteesMeetingMaterials/VeterinaryMedicine AdvisoryCommittee/UCM227005.pdf [accessed 2 May 2014].

engineered and non-genetically engineered Atlantic salmon were supplied by AquaBounty and the FDA also purchased Atlantic salmon from a local market to act as 'internal controls'.[97] Comparing Isoelectric Focusing (IEF) gel banding patterns (which separate different molecules, such as proteins, by differences in their isoelectric point) and 2-dimensional gel electrophoresis fingerprints (which separate DNA fragments based on their size) of AquAdvantage Salmon and non-genetically engineered Atlantic salmon, the FDA concluded that the essential nature of AquAdvantage's Salmon has not changed as a result of genetic engineering. In other words: AquAdvantage Salmon is an Atlantic salmon and there are 'no material differences in food from [AquAdvantage Salmon] … and other Atlantic salmon'.[98] Furthermore, even though there were 'slight gene expression changes' in AquAdvantage Salmon these were the result of natural variations.[99]

 After satisfying itself that AquAdvantage Salmon is an Atlantic salmon, the FDA had to be convinced that AquAdvantage Salmon is safe to eat. As we have already explained, in order to do this the FDA conducted comparative analysis (between AquAdvantage Salmon and relevant comparator Atlantic salmon) to determine whether AquAdvantage Salmon was 'substantially equivalent' to non-genetically engineered Atlantic salmon and that there was a 'reasonable certainty of no harm' from its consumption.[100] To satisfy itself, the FDA – informed by the earlier steps in the evaluation process as set out in *Guidance 187*, particularly the molecular characterisation of the construct and whether there are any added risks of consuming food containing the expression of the construct – characterised, and evaluated, any food consumption risks as either direct or indirect.[101] Direct risks are those risks associated directly with the consumption of the expressed inserted construct. Indirect risks are those risks not directly associated with the expression of the inserted construct but with changes in the composition or physiology of the genetically engineered animal such as nutritional deficiency and fat content. For both direct and indirect risks the assessment was based on comparisons between the genetically engineered animal and 'natural', non-genetically engineered comparators.

 As a starting point, the FDA views genetically engineered foods as safe to eat. In 1992, the FDA set out its policy for regulating biotechnology crops and concluded that genetically engineered crops were not substantially different from conventionally grown foods and were therefore safe to eat.[102] In terms of

---

97   FDA, above n 82, 64. It is noted that the comparative fish was largely farmed fish and therefore subject to more genetic modification through selective breeding than its wild counterpart.

98   Ibid., 109.

99   Ibid., 64.

100   Ibid., 61.

101   See FDA, above n 80, 23–24.

102   FDA, Federal Register, *Statement of policy: Foods derived from new plant varieties*, 57(104), 22984–92, 29 May 1992. Subsequent guidance on consultation between

specific direct hazards with AquAdvantage Salmon, the FDA identified particular risks associated with the expression of the Chinook salmon growth hormone and humans' potential allergenicity to the AquAdvantage Salmon. For example, the FDA identified that Chinook salmon, as finfish, are one of the eight main allergenic foods in the United States. But, the FDA recognised that while individuals allergic to Chinook salmon were also likely to be allergic to Atlantic salmon, these people would generally avoid all Atlantic salmon, including AquAdvantage Salmon.[103] So eating AquAdvantage Salmon does not pose any additional risks over and above the risks in 'natural', non-genetically engineered Atlantic salmon.[104]

In terms of indirect hazards, the FDA carried out various compositional analyses by comparing AquAdvantage Salmon with 'natural' non-genetically engineered Atlantic salmon. In so doing, the FDA assessed the level of numerous nutrients including carbohydrates, protein, fat, vitamins and minerals in AquAdvantage Salmon, as well as the endogenous allergenicity of AquAdvantage Salmon fillets, comparing them to one or more comparators. While the FDA requested supplemental data on the allergenicity of the tested tissue; they nonetheless concluded that:

> ... assessments of composition have determined that there are no material differences in food from [AquAdvantage Salmon] and other Atlantic salmon. We conclude that food from the [AquAdvantage Salmon] that is the subject of this application is as safe as food from conventional salmon, and that there is a reasonable certainty of no harm from consumption of food from [AquAdvantage Salmon].[105]

In 2010, after conducting its food safety assessment, the FDA concluded that AquAdvantage Salmon was safe to eat.

What we have shown is that the need to assess genetically engineered animals for 'substantial equivalence' requires a comparison with equivalent, non-genetically engineered animals. The food safety requirement of 'substantial equivalence' diverges somewhat from the requirement of United States patent law in which the USPTO has to be satisfied that genetically engineered animals have markedly different characteristics from something occurring in nature and are the result of human intervention. Indeed, the FDA application has the opposite focus, requiring the applicant to convince the FDA that genetically engineered animals are as close

---

GE-crop developers and FDA, (1997) Center for Food Safety and Applied Nutrition, *Guidance on consultation procedures: Foods derived from new plant varieties*, October 1997. [Online]. Available at: http://www.cfsan.fda.gov/~lrd/consulpr.html [accessed 2 May 2014].

103   FDA, above n 82, 75.

104   Ibid., 76. This was farmed Atlantic salmon. At the time, 90 per cent of the Atlantic salmon that was consumed in the United States was farmed: see VMAC, above n 96, 209.

105   FDA, above n 82, 109. The FDA did, however, request further study on the allergenicity of the tested tissue.

to 'natural' Atlantic salmon as possible. The FDA, for example, concluded that the essential nature of AquAdvantage Salmon has not changed as a result of genetic engineering and that any genetic variations were the result of natural variations.

Conclusion

Genetically engineered animals cannot be judged by a single standard. There are different ways of looking at genetically engineered animals, each being dependent on very specific legal, political, social, economic and practical conditions. By examining a particular instance of the paradox of nature found in arguments in favour of patenting and eating genetically engineered animals, this chapter has revealed how genetically engineered animals, such as AquAdvantage Salmon are presented in different legal contexts. More specifically this chapter has identified a paradox of nature in genetically engineered animals that is apparent in the way in which genetically engineered Atlantic salmon was presented to the USPTO and the FDA. Two judgments have, or will be, made on the same genetically engineered animal.

On the one hand, to satisfy the patentable subject matter requirement in patent law, genetically engineered animals must be shown to be 'markedly different' from similar animals occurring in nature. To do this, researchers and inventors of genetically engineered animals use images, descriptions and claims to show that the claimed genetically engineered animal (as a manufacture) is distinguished from, has 'markedly different' characteristics to and has been transformed into a different state to similar animals existing in nature. On the other hand, to be deemed safe for human consumption genetically engineered animals need to be 'substantially equivalent' to conventionally bred animals. In regards to AquAdvantage Salmon the FDA, for example, concluded that it looks and tastes like other Atlantic salmon, that the essential nature of AquAdvantage Salmon did not change as a result of genetic engineering, that any genetic variations were the result of natural variations and that there were no additional risks in consuming AquAdvantage Salmon.

While this chapter has focused on the paradox of nature, similar paradoxes may be identified and associated with issues of toxicity, safety and labelling. Nature and the natural – like toxicity, safety and ethics – exist in specific contexts and have different meanings depending on these contexts. So, when thinking about genetically engineered animals we need to be attentive to the different ways of validating genetically engineered animals. Indeed, the way in which genetically engineered animals are presented often reflects the particular legal, political, social or practical conditions in which the animal is being situated. It is necessary, therefore, to identify what is valued and ask a range of question including: How are genetically engineered animals evaluated and validated? What are the assumptions, propositions or underlying values contained in arguments over genetically engineered animals? What are the consequences of these propositions for human health, ecosystems and the granting of rights? In answering these questions, we cannot escape the fact that genetically engineered animals can be paradoxically both natural and something other than natural. In other words, they can be at the same time markedly different and yet substantially equivalent to animals found in nature.

Chapter 9

Information About Information About Information: GMOs and Law as a 'Flexible Technology'

Kieran Tranter

Introduction

It is not unusual for GMOs to be considered a prime exemplar of what Donna Haraway has termed 'natureculture'.[1] For many critics and commentators there is slippage at the essential level of categorisation when considering GMOs – living but not wholly natural, manufactured but without the industrial hallmarks usually associated with artefacts. Terms such as synthetic and hybrid are often deployed in critical accounts to capture this ambiguity. This chapter does little to resolve anxieties about GMOs. Rather it suggests that the categories that drive the anxiety have been erased, but not in some dialectical process where the third emerges from the antagonisms of the two. Instead, there has been a victory, possibly won many moons ago, of the manufactured over nature, of the commensurability of the all as manipulable information. Evidence for this claim is not to be found in the usual place. Rather than an examination of GMO discourse, or science, technology and society literature, the evidence for this argument lies in the mundane and routine practice of law.

Monsanto Company v. Syngenta Seeds Pty Ltd [2006] FCA 228, an application for preliminary discovery heard in the Federal Court of Australia before Justice Finkelstein, concerned, on its surface, none of these deep considerations. Monsanto was seeking to identify which entity within the Syngenta group had provided the Australian national scientific agency (Commonwealth Scientific and Industrial Research Organisation; CSIRO) with GM cotton seeds that Monsanto believed contained a patented gene. However, just beneath the 12 paragraphs concerning the application of several provisions from Order 15A of the then *Federal Court Rules*, there is a revelation concerning, pardon the pun, the nature of GMOs and the world they inhabit. The decision concerned information about information about information. It was information all the way down. Within this brief interlocutory,

---

1   Haraway, D. 1997. *Modest_Witness@Second_Millennium.FemaleMan©_Meets_ OncoMouse™: Feminism and technoscience.* New York: Routledge, 149.

procedural decision, the function of law as birthing a technologised world of manipulable information can be seen.

This chapter is in three sections. The first section orientates the chapter by reviewing anxieties about GMOs as a problem of hybridity. The second section reads *Monsanto Company v. Syngenta Seeds Pty Ltd* as a fundamental text. This involves multiple readings of the decision, a peeling back of the mundane proceduralism of rules and case authority to reveal, first the appearance of things – 'synthetic gene sequence', 'vip3A' gene and certain corporations – and second the dissolution of these phantoms into information blocks. The third section repositions this realisation concerning law as *Enframing*, a primal ordering of the world into manipulable information.

GMO as Hybridity

The emergence of GMOs has been a source of much anxiety within various literatures. For lawyers and policy makers GMOs represent a novel category, a new thing that potentially went beyond the reach of the existing law. GMOs were received as an invitation to legislation and the operative imagery has been the familiar legal anxieties of the law being 'outpaced' and needing to 'catch-up'.[2]

This is the default position adopted by the law when confronted with challenging technological change, the first move in what has been termed the 'law and technology enterprise'.[3] What goes mostly unsaid within the legal GMO literature is the *prima facie* assumption that GMOs are novel and challenging to existing law. GMOs are received into legal discourse as a problem in search of law. The lawyers' brief is not critically to assess GMOs as problematic, rather it is the practical task of determining frameworks and principles to balance the perceived rights and risks suggested by GMOs.[4] This legal discourse accepts the

---

2   Bennett Moses, L. 2007. Recurring dilemmas: The law's race to keep up with technological change. *Journal of Law, Technology and Policy*, 7(2), 239–85.

3   Tranter, K. 2011. The law and technology enterprise: Uncovering the template to legal scholarship on technology. *Law, Innovation and Technology*, 3(1), 31–83.

4   See, for examples, Hawker, N. 2013. Competition issues arising from generic biotech crops. *Drake Journal of Agricultural Law*, 18(1), 137–56; Helme, M. 2013. Genetically modified food fight: The FDA should step up to the regulatory plate so states do not cross the constitutional line. *Minnesota Law Review*, 98(1), 356–84; Wang, C. and Yu, W. 2012. Agro-GMO biosafety legislation in China: Current situation, challenges, and solutions. *Vermont Journal of Environmental Law*, 13(4), 865–84; Smith, L. 2012. Divided we fall: The shortcomings of the European Union's proposal for independent member states to regulate the cultivation of genetically modified organisms. *University of Pennsylvania Journal of International Law*, 33(3), 841–70; French, D. 2010. The regulation of genetically modified organisms and international law: A call for generality, in *The Regulation of Genetically Modified Organisms: Comparative Approaches*, edited by Bodiguel, L. and Cardwell, M. Oxford: Oxford University Press, 355–74; Pollack, M. and Shaffer, G. 2009.

assessment, made elsewhere – in 'society', in the media, by bioethicists, portrayed in science fiction – that GMOs are problematic.

The origin of this assessment is not hard to uncover. In *Our Posthuman Future*, Francis Fukuyama sees in biotechnology a worrying end. For Fukuyama biotechnology upsets an essential ordering of the world, the distinction between 'nature' and 'culture'.[5] He suggests that a healthy distinction between the wonder and bounty of the universe has existed since before humanity and before the products and artefacts that humans fashion from nature. This is not a radical move. It brings to the surface a fundamental distinction that many have identified within Western culture; the value laden dichotomy between 'nature' and 'culture' where nature is subservient to culture.[6] For Fukuyama biotechnology disrupts nature and culture in dangerous ways.[7] The distinction between natural and artefact, all so obvious with the big machines of industrial technology, has become blurred. There are theological and environmental traces behind Fukuyama's anxiety. Biotechnology is evidence of 'playing God' an arrogant overconfidence in human technical skill that carries risks to the biosphere that cannot be predicted. Fukuyama's book integrates seamlessly into the legal literature on GMOs. He presents the problem and goes on to suggest law as the solution,[8] paving the way for the legal literature on GMOs to catch law up and determine the best legal tools to reinforce the nature/culture divide.[9]

Other theorists are less interested in salvaging the nature/culture divide. For Donna Haraway GMOs are one in a long line of monsters to emerge from the implosion of the nature/culture divide. For Haraway, talk of nature or culture as separate categories that interact at specific sites according to specific known ways (the scientist making observations in the lab, the miner taking minerals from the earth) is, in this epoch of advanced technology, misleading. It obscures a deeper reality of dynamic interaction and hybridity that she, in the fabulously titled *Modest_Witness@Second_Millennium.FemaleMan©_Meets_OncoMouse™*, terms 'natureculture'.[10] GMOs as natureculture are living but not wholly natural, manufactured but without the industrial hallmarks usually associated with the artefact. In this GMOs are neither natural nor cultural but a hybrid of both. For Haraway, like Fukuyama, GMOs read as hybrids suggest the need for the making

---

When Cooperation Fails: The international law and politics of genetically modified foods. New York: Oxford University Press; Nelkin, D., Sands, P. and Stewart, R. 2000. The international challenge of genetically modified organism regulation. *New York University Environmental Law Journal*, 8(3), 523–9.

5   Fukuyama, F. 2002. *Our Posthuman Future: Consequences of the biotechnological revolution.* New York: Farrah, Straus and Giroux, 7, 97–100.

6   Feenberg, A. 1999. *Questioning Technology.* London: Routledge.

7   Fukuyama, above n 5, 84–102.

8   Ibid., 203–17.

9   Fukuyama suggests that GMOs, at least in agriculture, are one area where law and regulation has been proactive in managing risks: *ibid.*, 196–200.

10   Haraway, above n 1, 149.

of new, more appropriate and adapted laws. Unlike Fukuyama, Haraway is agnostic about GMOs as bits of natureculture. She does not directly suggest legal and regulative structures to restrain. Rather, she is concerned with understanding the complexities of the technoscience apparatus that thinks and generates GMOs so as to sketch the terrain in which new and more appropriate forms of politics and ethics can be established.

For some, this assessment of the hybrid newness of GMOs is misleading. Bruno Latour in *We Have Never Been Modern* establishes a nexus between modernity's drive to order and categorise and the unordered, disrespecting of these boundaries way of the world.[11] For Latour hybridity is 'natural'.[12] Modernity with its fields of knowledge and well-ordered world has been a conjurer's illusion; this knowing and ordering has never been complete. The very process of knowing and ordering squeezes out essential complexities that then return to compromise the knowing and ordering.[13] The hybridity of GMOs is nothing new. Within Latour's schema they are a recent monster whose hybridity shows the illusion of modernity. The difficulty and therefore the anxieties about GMOs for modernity highlight modernity's intellectual failings rather than anything new and challenging about GMOs.

But there is an interesting re-inscription of the nature/culture divide implicit in Latour's schema − a vitalist world of nature oozing from the pages of *We Have Never Been Modern*. This world that is continually confounding modernity with its hybrids suggests a premodern mythic nature, ultimately beyond human reason and control − a sort of unreasonable, that is unable to be reasoned, 'becoming' that disrupts and surprises human knowing and confounds human doing. Potentially this amounts to a Thomist metaphysics where there is an unknowable divine plan that can only be grasped incompletely through reason and natural law.[14] This would be exactly what it means never to have been modern. Traces of this approach of hybridity can be seen in other thinkers. Rosi Braidotti, although drawing from Haraway and Deleuze, rather than Latour specifically, champions this vitalist nature.[15] She reinvigorates the old term *zoe* as the name of this life that 'becomes' and surprises in contrast to the *bios* as known and ordered life.[16] *Zoe* is animalistic, bare life, while *bios* is the hybrid life that has been 'produced' by technoscience.[17] GMOs are *bios*. However, *zoe* is not entirely subdued in the processing of life to *bios*. The random, the chance, the accident, the mutation that is *zoe*'s essence

---

11   Latour, B. 1993. *We Have Never Been Modern*. Cambridge, Massachusetts: Harvard University Press.

12   Ibid., 40–42.

13   Ibid., 51.

14   Aquinas, T. 1988. Summa theologiae in *Saint Thomas Aquinas on Law, Morality and Politics*, edited by Baumgarth, W. and Regan, R. Cambridge: Avatar Books, 16–28.

15   Braidotti, R. 2006. *Transpositions: On nomadic ethics*. Cambridge: Polity Press, 37–41.

16   Ibid., 109–11.

17   Ibid., 55.

remains in *bios* – the weeds in cracks in the pavement, the gene that 'escapes' into wild populations. However, this seems to be a return. In attempts to get away from the nature/culture divide what can be seen is the re-inscription of a nature/culture divide that in form seems remarkably similar to the modern notions of subjectified nature and masterful culture. Braidotti's view identifies a realm of nature and a realm of culture and a processing of nature to culture. What has changed is the modern arrogance that nature is just a passive supplier of resources and culture as all knowing and all capable. Instead, nature's vitalism needs to be expected and respected and technoscience's limits emphasised. Notwithstanding, Braidotti's direct commitment to progressive politics and 'transpositioning', her ontological commitments potentially cast law in exactly the same sort of role as Fukuyama's call for it to humanise the posthuman future. A logical, possibly unintended, consequence of Braidotti's work is that law, on this view, is cast as a controlling device; regulating the unknown risks of vitalism and setting and reinforcing boundaries for technoscience.

What the preceding discussion shows is a tight triangle of GMOs-hybridity-law, with a surprisingly uncritical account of law. Law is presented, directly in Fukuyama and by implication in Braidotti, as having a master function. Law governs the nature/culture divide. It gives names and structures to blobs of 'nature' and polices technoscience's doings with these blobs. Taken together, Fukuyama, Haraway, Latour and Braidotti appear surprisingly uncritical in their accounts of law in that they pass over modern law's history as the West's other great order of life.[18] In doing so they all seem to suggest that the arrogant and dangerous values of modernity that had constructed colonialism, capitalism and patriarchy are not intrinsic to law's meta-function as master. Their revealed commonality is their perception of law as the order of the world and that this power is distinct from the values that law has legislated. This vision of law is ultra-positivist. Philippe Nonet has argued that 'positivism' ultimately names what law had become in modernity – a power that orders life – and that this power can be seen as separate and distinct from the historically contingent values that law imposed through its ordering.[19] Latour has recently begun to focus on modern law's function as 'formatter' of the world.[20] In *Aramis or the Love of Technology*, Latour made some preliminary observations about law as a 'flexible technology' that, through forms and texts and signatures, allows other technologies to 'harden'.[21] Law gets Latour's attention, not because it is a tool that political communities can use to regulate technoscience's hybrids, but because it is deeply enmeshed in technoscience's making of the world.

---

18   Fitzpatrick, P. 2001. *Modernism and the Grounds of Law.* Melbourne: Cambridge University Press.

19   Nonet, P. 1990. What is positive law. *Yale Law Journal*, 100(3), 667–99.

20   Latour, B. 2010. *The Making of Law: An ethnography of the Conseil d'Etat.* London: Polity, 268.

21   Latour, B. 1996. *Aramis or The Love of Technology.* Cambridge: Massachusetts: Harvard University Press, 45.

Law orders and structures actors and provides the protocols for their interactions. Latour suggests that law provides the basic coordination matrix, through which human and other resources are combined and deployed. From this perspective it is through law that technologies live.

From this legal perspective GMOs potentially take on a different characteristic. The view of GMOs as hybrids portrays GMOs as a technoscience object that slips between nature and culture, between *zoe* and *bios*. As has been discussed there is a concern with categories, with maintaining categories as in Fukuyama, disputing categories as in the Latour of *We Have Never Been Modern* or forging new categories as in Haraway and Braidotti. This emphasis on categories and categorisations suggests law as a primal order – that it is through law that categories are established, maintained and changed. However, as has been argued, law as primal order is suggested in this role but is not the focus. It opens the question of what do GMOs look like from this primal perspective. It means going in search of GMOs – not in the lab, nor in the texts of critics of technoscience – but in the law itself.

In Australia the obvious site for exploring GMOs in law is the *Gene Technology Act 2000* (Cth) and the regulative and administrative machinery that it establishes for the control of GMOs.[22] However, focusing on the GMOs that emerge from GMO law re-introduces the sort of categorising that Latour sees as misleading. Australian GMO law and regulation has several stories to tell about regulations, risk, the politics of technoscience and the tension between 'public' and 'expert' decision-making.[23] But it tells these stories from the technoscience perspective of GMOs as anxiety generating hybrids. GMO law tells us nothing about the other ways that law surrounds and orders GMOs. A better strategy for catching these elusive ordering relations lies in an examination of GMOs' entanglements with non-GMO law. *Monsanto Company v. Syngenta Seeds Pty Ltd* [2006] FCA 228, a decision of the Federal Court of Australia, is exactly a legal story of GMOs that is not about GMO law. Indeed, it seems very far removed from the heady politics and compromises of GMO law formation and operation documented by Richard Hindmarsh.[24] Formally, *Monsanto Company v. Syngenta Seeds* is a case concerning a mundane and procedural question of whether the court should make certain orders at a point prior to the formal commencement of litigation.

---

22 Tranter, M. 2003. A question of confidence: An appraisal of the operation of the Gene Technology Act 2000. *Environmental and Planning Law Journal*, 20(4), 245–60; Ludlow, K. 2004. Cultivating chaos: State responses to releases of genetically modified organisms. *Deakin Law Review*, 9(1), 1–39; Lawson, C. and Hindmarsh, R. 2006. Releasing GM cenola into the environment: Deconstructing a decision of the Gene Technology Regulator under the *Gene Technology Act 2000* (Cth). *Environmental and Planning Law Journal* 19(3), 22–59.

23 Hindmarsh, R. 2008. *Edging Towards BioUtopia: A new politics of reordering life and the democractic challenge.* Perth: University of Western Australia Press; Reynolds, R. 2009. Detoxification, displacement and deferral: The democratic failures of GM regulation. *Griffith Law Review*, 18(3), 753–77.

24 Ibid.

Monsanto Company v. Syngenta Seeds

Monsanto Company v. Syngenta Seeds is a brief interlocutory application for preliminary discovery. Preliminary discovery covers a range of orders set out in the then Order 15A of *Federal Court Rules 1979* (Cth)[25] allowing the court to make orders before the formal commencement of litigation. As a text the decision is brief, as is the norm for procedural law decisions. It comprises 12 paragraphs of reasons and has a total (headers and orders included) of 1,628 words. However, within this brief document there can be identified three distinct yet interconnected discourses about law and GMOs. The first is procedural, encapsulating the formal process whereby the civil procedure rules are applied to the facts. The second is technoscientific, referring to the various actors and their relations. The third is informatic, reflecting the way that the decision dissolves objects into information.

As a piece of procedural law, Justice Finkelstein's judgement is elegant and clear. The applicant, Monsanto, had asked the court to make orders for 'preliminary discovery' under Order 15A. Order 15A allowed orders to be made, outside of a commenced action, for parties to disclose information to the applicant. Specifically, Monsanto had requested from the court two orders. The first was an order for 'identity discovery' under Order 15A, rule 3. The second was an order for 'information discovery' under Order 15A, rule 6. The two orders were sequenced. Compliance with an order under rule 3 was required to ascertain the identity of the entity to which the rule 6 order should be addressed. Justice Finkelstein notes that the application began solely as an application under rule 6. The rule 3 request was added by amendment.This occurred when the named respondent in the unamended application, the Swiss based Syngenta Seeds Pty Ltd, was a stranger to Monsanto's complaint. This was accepted by Monsanto.[26] An order under rule 3 was requested to compel Syngenta to identify which of the entities within the family of Syngenta companies should be the subject of Monsanto's complaint.

Respondents subject to orders under O15A are potentially exposed to the significant costs of sorting through their archives to find the court-ordered information. It must be kept in mind that these orders are made before any formal litigation and for some respondents the application for the order could come as a complete surprise. The safeguard on the use of this power is that the court must be satisfied that 'there is a reasonable cause to believe that the applicant has or may have the right to obtain relief.'[27] This is a less exacting standard than the familiar

---

25 These are rules made under the *Federal Court of Australia Act 1976* (Cth) s 59. The *Federal Court Rules 1979* (Cth) that this decision was made under have since been repealed and replaced by the *Federal Court Rules 2011* (Cth).

26 *Monsanto Company v. Syngenta Seeds Pty Ltd* [2006] FCA 228, [4] (Finkelstein J).

27 *Federal Court Rules 1979* (Cth) O 15A, r 6(a).

'*prima facie* case' required for an injunction.[28] Nevertheless, it still required Monsanto to sketch the basis of its belief that it may have a right to obtain relief from an entity within the Syngenta group.

Monsanto's alleged cause of action was under section 117 of the *Patents Act 1990* (Cth). Section 117 makes suppliers liable for infringement if they provide a product to a person that, if used, would infringe a patent. Monsanto argued two allegations. The first, that Syngenta had developed cotton plants that infringed one of Monsanto's Australian patents for a 'synthetic plant gene'.[29] The second, an entity in the Syngenta group provided to the CSIRO plant material, most likely seeds, containing the contested synthetic plant gene.[30] The evidence base for the first allegation consisted of publications released by Syngenta that, in Monsanto's opinion, disclosed the breach.[31] The evidence base for the second allegation was information made public by the Office of the Gene Technology Regulator that the CSIRO had obtained at least four approvals to cultivate GM cotton that contained the contested genes and had, at the time of the decision, proceeded to trials according to two of the approvals.[32]

Syngenta did not present evidence countering these factual claims. Instead, the corporation argued for an interpretation of Order 15A that would prevent Monsanto from having the orders granted. It seems from the decision to be uncontroversial that in the circumstances set out in Syngenta's affidavit, if Monsanto did not have cause of action against it, an order under rule 6 could not be granted until a potential defendant was identified. This is implicit in the first limb of the rule 6 test of a 'reasonable basis of belief to obtain relief'.[33] One of the founding principles of litigation is that relief cannot be claimed against the world but must be claimed against individual named parties.[34] As Monsanto accepted it could not name which entities in the Syngenta group it believed it could obtain relief from, an order under rule 6 was unavailable on the facts as they stood.[35]

Having dealt with the rule 6 application Justice Finkelstein turned to the rule 3 application. Syngenta argued that an order under rule 3 must involve an assessment of the likelihood of the applicant 'succeeding in the action against the person whose identity is sought to be discovered'.[36] The case of *Hooper v. Kirella Pty Ltd* (1999) 96 FCR 1 was presented as authority.[37] Syngenta was attempting to make a chicken and egg argument. That as Monsanto was 'unable to identify

---

28   Cairns, B. 2014. *Australian Civil Procedure*. Sydney: Thomson Reuters, 441–2.
29   *Monsanto Company v. Syngenta Seeds*, above n 26, [1].
30   Ibid., [3], [5].
31   Ibid., [3].
32   Ibid., [5].
33   *Federal Court Rules 1979* (Cth) O 15A, r 6(a).
34   Cairns, above n 28, 118.
35   *Monsanto Company v. Syngenta Seeds*, above n 26, [4].
36   Ibid., [8].
37   Ibid.

a prospective defendant, it is also presently unable to articulate a cause of action against that unknown defendant. Accordingly, so the argument goes, Monsanto was simply not entitled to an order.'[38]

Justice Finkelstein rejected this interpretation of rule 3. He noted that the requirement of identifying a reasonable cause of action, as rule 6, is not in the wording of rule 3.[39] He also noted that, while cases like *Hooper v. Kirella Pty Ltd* do suggest that an assessment of the basis of the applicant's cause of action is often a useful safeguard when considering granting a rule 3 order, there is no 'absolute rule' requiring such an assessment. Furthermore, he suggested that such a 'condition' could give rise to injustice in cases exactly like the one before him.[40] Indeed, for Justice Finkelstein the reasons were 'so obvious they need no further elaboration'.[41]

Having rejected Syngenta's interpretation of rule 3, Justice Finkelstein went on to fashion the 'criterion' required for an order under rule 3. He determined that the appropriate test should require Monsanto to show that it 'has a claim that is worth investigating'.[42] On the basis of this evidence Justice Finkelstein concluded that Monsanto had a claim worth investigating. Indeed, the brevity of the decision is justified by statement that there was 'no need to spend much time considering whether Monsanto has a claim worth investigating'.[43] He described there being a 'reasonable basis for believing that Monsanto's patent may have been infringed' and that 'it seems clear enough that some company in the Syngenta group is involved'.[44] Following these findings an order for identity discovery was made and the order for information discovery was stood over.[45]

Monsanto 1: Syngenta 0. But that is where the procedural reading of the decision runs out. It is a piece of minor authority for the proposition that, in circumstances where there is a fog of related entities that an applicant needs to dispel in order to see which exact entity they might have a cause of action against, identity discovery is available.[46] However, the strength of it as authority might have been weakened by the new 'identity discovery' provisions in rule 7.22. The new provisions now include the term 'prospective respondent', defined as 'a person not presently a party ... [from] whom a prospective applicant reasonably believes the prospective applicant may have a right to obtain relief'[47] and directly states

---

38 Ibid.
39 Ibid., [9].
40 Ibid.
41 Ibid.
42 Ibid., [10].
43 Ibid., [11].
44 Ibid.
45 Ibid., [12].
46 Condon, W. and Polites, D. 2006. Preliminary discovery in the Federal Court of Australia – A practical approach. *IPL Newsletter*, 24(4), 21–3.
47 *Federal Court Rules 2011* (Cth) r 7.21.

the need for 'the Court to be satisfied that there may be a right for the prospective applicant to obtain relief against a prospective respondent'.[48] These inclusions in the new rules seem to weaken some of Justice Finkelstein's arguments about the application of identity discovery in a fog of related entities. They seem to suggest the need for more resolution as to the potential parties than is covered by Justice Finkelstein's 'claim worth investigating'. In this light the decision could be seen as a withered un-nurtured planting of procedural law – abandoned before it took hold – standing only for the substantive decision itself that Monsanto was granted identity discovery. But even at this level the decision is withered and un-nurtured. There appears to have been no further litigation on this matter. Notwithstanding the outcome of the identity discovery order, there does not seem to be any publically available decision on either the stood over information discovery application or any proper litigation arising from these investigations. This lack of subsequent litigation seems at odds with the representation of the parties in the decision. In bringing the preliminary application Monsanto emerges from the text as an aggressive defender of its intellectual property. Syngenta with its affidavit of denial and with its novel interpretation of rule 3 emerges as a stubborn, uncooperative respondent prepared to put whatever obstacles it can in Monsanto's way. The makings of happy litigation lawyers are suggested by these parties' attitudes. Yet the expected stack of subsequent decisions and judgements on this matter has not materialised. The parties skirmished but then seemingly abandoned the battle. The other intriguing aspect of *Monsanto Company v. Syngenta Seeds* and its lack of a legal heritage is that Monsanto was not chasing an order against the CSIRO. Yet the decision is clear that it was the CSIRO that Monsanto believed was the primary infringer of their patent. It was the CSIRO that was given approvals by the Office of the Gene Technology Regulator and it was the CSIRO that had actually conducted trials with GM cotton that Monsanto believed infringed its patent. The relationship between Monsanto and the CSIRO is documented in the decision. Justice Finkelstein reported that Monsanto requested information from the CSIRO and that the CSIRO responded that the information sought was subject to confidentiality agreements that required the consent of the other party to disclose and that the other party had refused to consent.[49] It seems uncharacteristic that an aggressive applicant like Monsanto would not seek an order against an obvious respondent like the CSIRO.

These intriguing oddities reveal the limits of reading *Monsanto Company v. Syngenta Seeds* as purely a matter of procedural law. There is clearly a wider context and that wider context is the technoscientific realm of science, politics and agribusiness. From a technoscientific perspective a web of 'actors' and their relationships interface across the decision. The starting point is the very actor that lies at the nexus of the relationship webs documented in the decision – what

---

48   Ibid., r 7.22(1)(a).
49   *Monsanto Company v. Syngenta Seeds*, above n 26, [6].

Finkelstein obliquely calls a 'synthetic plant gene'.[50] Technoscience names it more fully. It is the genetic sequence that can produce a 'vegetative insecticidal protein' (vip). Presence of a vip in a plant makes the plant more resistant to insect attacks. The specific vip under consideration in the decision was 'vip3A'. It has been described in the scientific literature as affecting selective insect herbivores through damaging the midgut lining.[51] Suddenly, a clearer backstory to the decision emerges. Monsanto holds Australian patents for a sequence that can produce vip3A and the material from the Office of the Gene Technology Regulator indicated that the CSIRO had been given approval to trial modified cotton plants that allegedly produced vip3A, and that the CSIRO had not received these from Monsanto.

Of course technoscience cannot solely be seen as just science as encapsulated by the Popperian vision of hypothesis, experiment, refutation and dissemination.[52] But there are traces of these practices within the decision. Monsanto referred to scientific publications concerning Syngenta's own research into vips. Further, the CSIRO's involvement suggests science. Its approvals are for 'trials'[53] and not for commercial cropping. However, it is commercial cropping and the mega-profits to be made from agriculture that are the technoscience story of the decision. In any thinking and writing about GMOs Monsanto and Syngenta loom large. They are the giant transnational composite entities engaged in fierce global competition for the world's food dollar.[54] This competition occurs in the realm of scientific research and technological development, but the focus is not on knowledge as in the Popperian model, it is much more on who can capture a market by inaugurating the next phase in the 'Green Revolution' and then defend that monopoly against the other.[55] This is the reality of technoscience in the decision; a complex of business meets science, profit meets innovation, global meets local.

Monsanto Company v. Syngenta Seeds from this perspective can be seen as a minor local engagement within a global war. Monsanto sought to 'own' the Australian market for vip3A expressing crops through patenting a genetic sequence that, if spliced into a plant's genome, resulted in it producing vip3A. The decision, in its reference to Syngenta's publications, seems to suggest that Syngenta had also developed a genetic sequence that produced vip3A but Monsanto believed it

---

50   Ibid., [1].

51   Lee, M., Walters, F., Hart, H., Palekar, N. and Chen, J.-S. 2003. The mode of action of the *Bacillus Thuringiensis* vegetative insecticidal orotein Vip3A differs from that of Cry1Ab δ-endotoxin. *Applied Environmental Microbiology*, 69(8), 4648–57, 4648.

52   Compare Popper, K. 1989. *Conjectures and Refutations: the Growth of Scientific Knowledge*. London: Routledge, 240–42 with Haraway, above n 1, 50–51.

53   *Monsanto Company v. Syngenta Seeds*, above n 26, [3].

54   See, for examples, Pringle, P. 2003. *Food, Inc: Mendel to Monsanto – The Promises and Perils of the Biotech Harvest*. New York: Simon and Schuster; Coburn, C. 2005. Out of the petri dish and back to the people: A cultural approach to GMO policy. *Wisconsin International Law Journal*, 23(2), 283–320.

55   Stumo, M. 2010. Anticompetitive tactics in ag biotech could stifle entrance of generic traits. *Drake Journal of Agricultural Law*, 15(1), 137–48.

had claimed the Australian market through the patent on its sequence. And Justice Finkelstein thought that there was enough to its belief for Monsanto to have 'a claim worth investigating' against an entity or entities within the Syngenta group.

In returning to the legal engagement between Monsanta and Syngenta a plethora of legal actors and relationships can further be identified. Foremost, is the patent itself. It is not described in the decision, but frames the entire document through Justice Finkelstein's establishing it in the opening line. And where there is a patent there is patent law. The *Patents Act 1990* (Cth), specifically s 117 dealing with supplier liability, appears in paragraph 7 to name Monsanto's possible cause of action. These are not the only 'law' actors weaving through the text. GMO law is present. Both the *Gene Technology Act 2000* (Cth) and the Office of Gene Technology Regulator are active. The Act is named as having birthed the four approvals granted to the CSIRO and the Office of Gene Technology Regulator is recorded as making publically available information about these approvals.[56] There are yet more legal actors in the decision. The rule 3 application was made precisely because it is entirely permissible within company law to have commercial activity conducted through a complex fog of interrelated corporations. Syngenta as a pluralistic and shifting target pervades the decision. There are also notions of contract and confidentiality provisions in relation to the CSIRO's denial of Monsanto's informal request for further information.

From this perspective various actors have been identified radiating out from a primary technoscientific thing – a modified genetic sequence that expresses a vip. From sequence comes the specific vip, vip3A, and from vip3A come traces of the scientific method and from the method emerge the well-known agribusinesses of Monsanto and Syngenta and their intense competition for profit in global agriculture. From this profit comes patents and from patents and patent law came a string of legal actors, from Australian GMO law and the regulatory body and approvals that it creates, to deeper more mundane legalities concerning the malleability of the corporate form and contracts with confidentiality provisions.

This technoscience perspective explains some of the anomalies that were identified after the procedural reading. The absence of any subsequent decision can be explained in the context of the titanic, global struggle between Monsanto and Syngenta. Like giant anime fighting robots, the combatants merely moved on to other theatres of engagement. The Australian application for preliminary discovery could have been a feint within a much more complex set of global manoeuvrings. Charles Lawson, in this volume, explains how Monsanto and other global agribusinesses use preliminary and interlocutory litigation as a precursor to negotiations leading to licencing agreements.[57] This global battle also explains Monsanto's leaving the CSIRO out of the application. The CSIRO is a small but significant producer of agricultural knowledge and is often the technical gatekeeper for entry of new agricultural 'products' into Australia. It is an entity that Monsanto

---

56   *Monsanto Company v. Syngenta Seeds*, above n 26, [3] and [5].

57   See Lawson's chapter in this volume.

deals with and will want to continue to deal with into the future. Whatever the feeling that could be ascribed to Monsanto, as sketched in the decision concerning alleged breaches of its Australian patent, it did not justify jeopardising relationships with the CSIRO by including it as a respondent.

There is also something to be said about how the technoscience perspective traced the legal. It might have started with the hybrid 'synthetic plant gene' but it ended with a minor cartography of law actors. This progression could be read in two ways. The first is a quirk of the examined text. A judgment by the Federal Court of Australia is going to be populated by law actors. It would not pass a threshold 'rule of recognition' if it did not conform to the expectations of judicial decision and the protocols of legal writing.[58] To be 'accepted' its intertextuality needs to involve case law and legislation and its flow needs to follow the principles of legal reasoning. To observe that a law text concerning GMOs talks of law is potentially meaningless. It could be expected that a scientific text on GMOs would talk of the conventions of science, of experiments, results and past studies, or a corporate text on GMOs by Monsanto or Syngenta would talk of productivity, yields and risk minimisation through bullet points and glossy charts. A second, more generous way, would involve connecting with Latour's later work on law as facilitating technoscience. The decision shows that technoscience and its hybrids do not come about just through the institutional workings of science or the institutional machinations of advanced capitalism but through a framework of law that structures relationships, allowing the production of knowledge, 'things' and wealth. *Monsanto Company v. Syngenta Seeds* directly brings to the surface elements of this framework. There is property signified by patents and patent law. There is regulation signified by the *Gene Technology Act 2000* (Cth), the regulator and the approvals. There is, in the fog of entities that travelled under the name of 'Syngenta,' the corporate form as the shadowy and shifting location of advanced capitalism. And there is contract with its features of privity and confidentiality associated with the CSIRO. The decision clearly substantiates the idea that technoscience lives through law. Technoscience is about scientific knowledge and about capital but these 'become' in the world through property, regulation, entities and contract. In this the decision exposes law as 'formatting', and GMOs as being and living through this primal ordering by law.

Having exposed the primal ordering by law, the next suggestion is how does law facilitate this ordering? This process of ordering can also be seen in *Monsanto Company v. Syngenta Seeds* although it is not revealed through the formal procedural perspective or the technoscience perspective but through a more exacting focus on the text of the decision. The decision is remarkable in its confidences and its ambit. It captures a plethora of entities and stiches them together. The sequence, vip3A, Monsanto, Syngenta, CSIRO, Office of Gene Technology Regulator, a patent, legislation, case law, a contract, procedural rules and orders are brought together according to the internal protocols of acceptable

---

58   Hart, H. 1961. *The Concept of Law.* Oxford: Clarendon Press, 97.

decision. All are captured in the text, written and organised. Law does not seem to suffer from what Latour identified as the myopic boundary policing of modern disciplines.[59] *Monsanto Company v. Syngenta Seeds* bundles together an iconic blob of natureculture that is the vip3A expressing genetic sequence along with other entities such as corporations, regulators and research organisations and more classic legal forms such as patents, contracts and case law. It is this knowing no boundaries, of being able to bring anything into law and mix it with anything else that Latour identifies as the uniqueness of law.[60] *Monsanto Company v. Syngenta Seeds* achieves this through the classic dark magic of naming and, through naming, controlling. All the entities that are ordered together in the decision are *named*. They are signified in the text as 'vip3A', 'Monsanto', 'patent'. Even entities, such as GMOs or legal entities within a corporate fog that are considered ambiguous and shifting are effortlessly named 'synthetic plant gene' and 'Syngenta'. In being named in the text of the decision they become ordered in time. There is an ordering of the past: this entity did this at this time and it had this effect. The CSIRO refused to release to Monsanto any further information. There is also a future ordering: orders are made directing a future set of connections between Monsanto, Syngenta, vip3A and the Court.

The decision ultimately concerns information, about information, about information. On the surface this is what Monsanto was asking the court to order; information about Syngenta's corporate structure, so it could get information about Syngenta's vip3A programme, so it could determine whether there is information that suggests there have been breaches of its patents. But at an essential level the decision discloses a worldview where all and everything is manipulable data. All there is, is information that can be named and organised. In law anything can be dissolved into a tight signification. Anxiety generating hybrids like GMOs become an information block – a 'synthetic plant gene' – that can be connected with other information blocks. Furthermore, this connecting is not just the writing of a history and fixing the record of past connections. Nor is it just writing for writings sake. Law as formatter means that this process connects blocks for the future. It makes the world.

Law as Enframing

Where does this leave GMOs? Translated into the language of category and hybridity that often accompanies discussion of GMOs, *Monsanto Company v. Syngenta Seeds* shows the illusion of such talk. It suggests that law dissolved the nature/culture divide many moons ago. There is no vitalist, un-representable nature in law's sovereign writing of the world. No category beyond the cultural to limit or confound or confuse. Law makes the world through naming and ordering.

---

59   Latour, above n 11, 10–11.
60   Latour, above n 20, 268–9.

There is something highly familiar to this claim. Martin Heidegger's 'The Question Concerning Technology' is often considered a starting point for contemporary critical approaches to technoscience. Heidegger perceived in technology a corruption of the human impulse to create.[61] It was a corruption, for instead of adding to the beauty of the world, it has reordered the world as standing reserve, as resources ready at hand to be deployed to whatever end.[62] This he called *Enframing*.[63] Modern technology or more precisely the worldview that births modern technology, does so having turned the world into information blocks. For Heidegger, the villain in this story of a fall is Western metaphysics since the classical Greeks.[64] What Heidegger leaves out are the processes whereby Western metaphysics becomes a 'hydroelectric plant set into the current of the Rhine' or 'an airliner that stands on the runway'.[65]

Monsanto Company v. Syngenta Seeds as a text might be light-years removed from Heidegger's 'The Question of Technology', yet there is a remarkable coherence between the two. *Monsanto Company v. Syngenta Seeds* shows law as *Enframing*. In its brief, mundane proceduralism it shows how 'things' become named and ordered in time. It demonstrates Latour's aside claim of law as the primal 'flexible technologies' through which other technologies, vip3A for example, 'harden'.[66] There are some implications that flow from this realisation.

First, in the realm of legal theory what is suggested is a much more promising path for 'Heidegger and law' than the existing small literature that has attempted to draw insights from Heidegger to law through often tortuous accounts of interpretation.[67] Second and more directly related to GMOs, this realisation emphasises the lack of anxiety that the machinery of law has had with GMOs. The information block 'synthetic plant gene' in *Monsanto Company v. Syngenta Seeds* could have been interchangeable with any other, less anxiety generating thing. The decision would have read much the same if. instead of synthetic plant gene and vip3A, it had concerned 'impact absorbing bumper bars' or 'non-leak fountain pens'. Also glimpsed in the decision is the entirely unremarkable regulatory legislation that establishes the Gene Technology Regulator and the licensing/approval scheme. This act, through criminalising certain conduct, setting up an executive decision maker and providing for a scheme of licencing and approvals,

---

61   Heidegger, M. 1977. *The Question Concerning Technology and other Essays.* New York: Harper and Row, 24.

62   Ibid., 26–7.

63   Ibid., 20.

64   Ibid., 34.

65   Ibid., 16–17.

66   Latour, above n 21, 45.

67   Crosby, I. 1998. Worlds in stone: Gadamer, Heidegger and originalism. *Texas Law Review*, 76(4), 849–67; Leiter, B. 1996. Heidegger and the theory of adjudication. *Yale Law Journal*, 106(2), 253–82; Scheibler, I. 2000. Gadamer, Heidegger and the social dimensions of language: Reflections on the critical potential of hermeneutical philosophy. *Chicago-Kent Law Review*, 76(2), 853–92.

is no different in form to much earlier legislative responses to other technologies.[68] Law was unperturbed by the anxieties surrounding GMOs. GMOs are just another information block to be named and ordered.

This second implication leads to the third implication, a reassessment of law. In Fukuyama and other critical accounts of technoscience, law is an end at the end. It is invoked to protect categories that technoscience is threatening. Law saves. What *Monsanto Company v. Syngenta Seeds* shows is that law is not innocent in technoscience's creation of troublesome hybrids. Instead, law prefigures technoscience's doing in the world. It is law's capacity to name and order and in that naming and ordering there is the establishing of future relations that allow technoscience to live. In Heidegger's schema law can be seen as the 'missing link' between Western metaphysics and technology; the machine behind the machine. It is law that genesis-es the technoscience world of GMOs – or less biblically – law 'Enframes' the protozoan mud pool from which our world emerges.

---

68   Tranter, K. 2005. The history of the haste-wagons: The *Motor Car Act 1909* (Vic), emergent technology and the call for law. *Melbourne University Law Review*, 29(3), 843–79.

Index